Mind, Matter, Cosmos, and Consciousness in Plain English

Why Atheism Is Wrong
About Our Immaterial Universe and
How the Mind Exists Beyond the Brain

Christoph Mlinarchik

About the Author, Christoph Mlinarchik

Christoph Mlinarchik is the author of Christianity in Plain English, the first book in the Deep Questions in Plain English Series, and he also wrote bestselling business books like Government Contracts in Plain English, Federal Acquisition Regulation in Plain English, and Government Contracts Negotiation, Simplified.

Christoph Mlinarchik owns the company Christoph LLC that provides expert advice in government contracts: consulting, expert witness services, and professional training. Contact Christoph LLC for media, interview, podcast, public speaking, book-signing, discounted bulk book purchasing, or business inquiries: **Christoph@ChristophLLC.com**

Owner Christoph Mlinarchik, JD, CFCM, PMP (Certified Federal Contract Manager, Project Management Professional) is an attorney, expert witness, professional instructor, consultant, frequent public speaker, nationally recognized subject matter expert, and award-winning author of 75+ publications on government contracts and acquisitions, including the three books in The Government Contracts in Plain English Series: **https://www.amazon.com/dp/B09MRCMWBD?binding=paperback** and also Christianity in Plain English: **https://www.amazon.com/dp/1734198109**

Mind, Matter, Cosmos, and Consciousness in Plain English: Why Atheism Is Wrong About Our Immaterial Universe and How the Mind Exists Beyond the Brain

www.ChristophLLC.com
Christoph@ChristophLLC.com

ISBN: 978-1-7341981-4-0 (paperback) 978-1-7341981-6-4 (electronic version)

This book explores provocative ideas within the spirit of intellectual inquiry, debate, and critical thinking. Exploration, explanation, or critique of ideas is not the same as endorsement or adoption thereof.

Table of Contents

Part 1, Introductory Ideas

Part 2, Matter and Cosmos

PART 3, MIND AND CONSCIOUSNESS

PART 1, INTRODUCTORY IDEAS

1 What to Expect in Mind, Matter, Cosmos, and Consciousness in Plain English

What type of answers will you read in this book?

Short, concise, direct answers to deep questions in plain English. My book is simple by design—a quick read to save you time and explain difficult concepts easily.

What is the purpose of this book?

Mind, Matter, Cosmos, and Consciousness in Plain English explores the immaterial aspects of reality as revealed by our minds and our entire universe.

How is this book organized?

After five introductory chapters (Part 1), the first half of the book (Part 2, Matter and Cosmos) highlights why a strictly materialistic outlook cannot explain our physical cosmos: the sudden beginning event of the Big Bang, the fine-tuned properties of the physical laws of the universe, and the immaterial nature of quantum physics. The second half of the book (Part 3, Mind and Consciousness) explores free will, consciousness, the mind-brain connection, and innovative neuroscientific research to confirm the immaterial properties of the human mind. After learning more about the immaterial nature of our minds and of our universe, my hope is that you will firmly reject the deluded, dangerous, and depressing worldview of strict materialism.

What is the worldview of materialism?

Materialism is the belief that all existence consists of nothing more than material objects. By reducing our entire universe and our minds to atoms and molecules crashing about, strict materialism conflicts with the reality of the immaterial concepts most important to humanity.

What is your first loss if you follow the materialist worldview?

If you believe in the worldview of materialism, your first loss is your very identity: your consciousness is simply an illusion created by the lump of brain tissue in your skull. Combining atheism with materialism makes things even more grim. If God does not exist, there is no objective morality, meaning, or purpose. We live and die for no reason with no hope of salvation, redemption, or transcendence. How depressing!

What aspects of our universe refute the materialist worldview?

Fortunately, the best evidence and reasoning refute the worldview of atheistic materialism. Our universe reveals scores of physical properties that point to fine-tuned design, eliminating the possibility that the universe emerged by chance or necessity. From the beginning, our cosmos is incompatible with atheistic materialism.

Scientific investigation confirms that some 13.8 billion years ago during the Big Bang event, all matter, energy, time, and space appeared from a state of absolute nothingness. Because the Big Bang marks the first appearance of matter itself, the worldview of materialism is impotent to explain the beginning of the entire universe. With no matter before the Big Bang, no material explanation is possible. Even more intriguing, quantum physics reveals that the fundamental properties of physical matter itself point to immaterial, abstract, mathematical probabilities that are intelligible only to immaterial minds.

What aspects of our minds refute the materialist worldview?

Turning to the human mind, cutting-edge neuroscientific research supports immaterial concepts like free will and consciousness, thereby rejecting key assumptions of atheistic materialism. Groundbreaking neuroscientific experiments follow materialist procedures yet surprisingly confirm immaterial aspects of the human mind like indivisibility and nonlocality. By reading about neuroscientific exploration into seizures, memory, free will, involuntary brain stimulation, split-brain surgery, conjoined twins, and even Near-Death Experiences, you will realize that the atheistic materialist worldview fails to account for a wide range of immaterial mental activity in your daily life. Therefore, the atheistic materialist assumption that the mind is nothing but the physical brain is wrong. The immaterial mind exists beyond the three-pound organ in your skull.

How will this book improve your thinking, understanding, and outlook?

After reading this book, you will understand the reasoning behind the spirited debate over the nature of our reality, our cosmos, and our minds. While I hope to lead you closer to my point of view, my paramount goal is to equip you with the knowledge, evidence, and reasoning—in plain English—to make intelligent, informed decisions about the greatest mysteries of reality.

2 Is "Trust the Science" Self-Defeating? Scientism and Atheistic Materialism Masquerading as Science

Scientism – distinct from science – is a philosophical belief that truth is found only within scientific experimentation. Following scientism means that concepts like love, God, relationships, meaning, and purpose contain no truth and are fundamentally invalid, simply because they cannot be confirmed by laboratory tests. Paradoxically, science itself cannot prove its own validity because science relies on non-scientific assumptions, like mathematics and logic, which cannot be proven through scientific experimentation. Therefore, scientism is not only contrary to the full range of human experience, but also hopelessly self-refuting and in conflict with entire categories of essential truth.

Atheistic materialism is a tragic worldview that denies God, souls, spirits, and even human consciousness and the mind beyond the brain. Limiting our entire universe and existence to material causes, atheistic materialism cannot account for either the origin of the cosmos (before the Big Bang) or the origin of our minds. Our existential beginning and daily existence are completely out of reach for physical or materialist explanations because all matter appeared suddenly during the Big Bang and because the mind is immaterial yet interacts constantly with our material body. While other worldviews open the door to understanding the mysteries of the mind and cosmos, atheistic materialism closes the door. Unless you abandon atheistic materialism and its prejudicial worldview focused exclusively on the physical, you will be constantly disappointed with unreasonable, shallow, and silly excuses for immaterial mysteries like the origin of the mind and cosmos.

How does scientism relate to scientific experimentation?

Scientism is the belief that no truth exists except that which is proven by scientific experimentation. Notably, belief in scientism means disbelief in God, souls, spirits, objective morality, purpose in human lives, meaning in the universe, and a lengthy list of other essential truths which cannot be proven in a laboratory. Therefore, scientism is an incredibly narrow and sterile worldview.

What is science? What is the etymology of the word?

"Science" is a controversial word like democracy, justice, or art. Due to so many competing and conflicting claims on its meaning, the intellectual treasury of the word "science" is massively and hopelessly overdrawn and insolvent. Science is a bankrupt term and cannot possibly pay all its intellectual debts to its countless claimants.

The word "science" comes from the Latin word, "scientia," meaning "knowledge." The Latin "scientia" is the translation of the original Greek word for structured knowledge or understanding, "episteme."

What is one elegant, practical definition of science?

One elegant, practical definition of science is the systematic study of causes and effects in the cosmos, universe, or natural world. Note that in this definition of science, the types of causes and effects to study are limitless, with no preconceived biases or foregone conclusions.

Has modern science excluded entire categories of knowledge from scientific study?

Yes. Throughout the broad history of science, subjects like theology, metaphysics, and philosophy were included within the scope of science. In modern times, the common conception of science has narrowed significantly by excluding any immaterial or supernatural causes, and by ignoring the intellectual foundations of science, namely, theology and philosophy (see chapter 4, "The Scientific Mind and its Dependence on Philosophy, Theology, and Immaterial Concepts"). Nowadays science has categorically outlawed the study or consideration of anything outside the narrow, myopic worldview of atheistic materialism. For the fundamental mysteries of the origin of mind, life, and cosmos, modern science is cornered into empty, sterile excuses and obfuscation when the evidence points to immaterial, supernatural causes.

What is atheistic materialism? What does atheistic materialism exclude?

Atheistic materialism is a worldview that denies God and anything immaterial, instead believing only in material causes and phenomena. If you are an atheistic materialist, you deny not only God but also souls, spirits, immaterial minds, the afterlife, and anything religious. Atheistic materialism will not accept any supernatural or immaterial explanations no matter the evidence, which creates many awkward and unconvincing "explanations" of physical causes for non-physical phenomena like the origin of the entire cosmos, or the existence of consciousness, mind, and soul.

What is one obvious reason why immaterial or supernatural causes must be considered within (not excluded from) scientific exploration? What has science discovered about the beginning of the cosmos?

Our cosmos consists of the entire natural universe, and its beginning cannot be explained by material causes. Scientific research has firmly established the fact of the Big Bang, which is the beginning of all time, space, matter, and energy in our cosmos. If you limit your investigation to atheistic materialism, you cannot explore non-material or supernatural causes.

Yet the Big Bang itself—the mystery of the beginning of time itself, the entire material cosmos, and therefore everything we explore in science—necessarily requires a non-material, supernatural cause. Before the Big Bang, there was no matter, energy, space, nor time to cause the Big Bang. Therefore, the cause of the Big Bang cannot be derived from a material cause, because matter itself did not exist until it somehow came into being, from nothing, in the Singularity event of the Big Bang some 13.8 billion years ago. See chapter 7, "The Beginning of Everything: The Big Bang."

What is the scientific method?

The scientific method is a systematic, incremental approach to discovering knowledge. The scientific method is systematic because it follows well-defined processes, and it is incremental because it continues indefinitely as new findings provoke new questions and further exploration. The process of the scientific method can be summarized in six steps:

1. Observation. Record and summarize what you sense and ascertain about whatever subject you pursue. You can observe the migration path of birds, or the melting of ice, or the growth of plants in your backyard.
2. Forming a hypothesis. When you form a hypothesis, you attempt to explain your observations, or to formulate some relationship, principle, rule, or other insight.
3. Prediction. Based on your hypothesis, you predict what will happen in the future or during experimentation. Your

experiment will be designed to test your prediction and therefore assess the hypothesis that explains the prediction.

4. Experimentation. Perhaps the most familiar part of the scientific method is the experiment, whereby you test your hypothesis in a controlled setting. One key ingredient of a successful experiment is the so-called "control" group. If your experiment tests the hypothesis that a new plant food additive will accelerate growth, your "control" group would be plants in the same setting that do not receive the food additive. You would measure the growth of plants that receive the additive and compare those results to the plants in the control group. Any differences may reveal the effect(s) of the food additive on plant growth.
5. Analysis. After experimentation, the scientific method invites rational analysis of the results. Careful measurement is important, as is the exclusion of any data that is corrupted by improper procedure.
6. Conclusions. Ideally, your analysis of the experimentation results will yield certain conclusions, although this will not happen in every experiment. Some experiments yield contradictory, muddled, or incoherent results and therefore escape firm conclusions. But if your analysis reveals clear conclusions, the scientific process has not ended, it has merely begun. Firm conclusions often pose further questions and hypotheses that are ripe for further experimentation, and then the scientific process continues in its characteristically incremental process. Back to step one!

What is scientism? What is the difference between science and scientism?

Scientism is the narrow worldview that truth is found only within physical, material, scientific experiments. Nothing is true other than what can be proven in a laboratory or by the scientific method. Science is the only source of truth if you believe in scientism.

Scientism is a perversion of science into a quasi-religious worldview. Scientism is an ideological belief that denies any truth except that which is provable through scientific experimentation. The problem with scientism is that most scientific experiments are constrained to investigate only material phenomenon in the natural world with little bearing on the most important subjects for humanity like morality, good and evil, beauty, political and social conventions, and even truth itself. Further, science cannot prove the assumptions and conventions that make science itself possible, like formal logic and mathematics. Scientism is therefore self-refuting and internally inconsistent.

Science, on the other hand, is the systematic study of causes and effects, including but not limited to the structured activity known as the scientific method. Open scientific inquiry explores all knowledge, all causes, and all effects—regardless of ideological, philosophical, or theological beliefs or worldviews.

In contrast to science, scientism is an ideological belief system and epistemological (truth) claim, not a method of searching for truth. Scientism is an exclusionary principle that only knowledge and truth proven through the scientific method is valid, and anything else should be ignored or rejected. When you consider how many essential truths and concepts cannot be proven in a laboratory, the intellectual poverty of scientism becomes glaring.

What is methodological naturalism? What is scientific naturalism? How do scientific naturalism and methodological naturalism relate to scientism?

Methodological naturalism is the practice or process of excluding every explanation except for the material when conducting scientific experiments. The methodological naturalist considers only physical, material causes when analyzing experimentation results and carefully excludes any other immaterial explanations. The connection to scientism is obvious.

Under the limited worldview of scientism, no truth outside the bounds of scientific research investigating material causes is valid. Following methodological naturalism means that when conducting scientific research and exploration, only physical, material causes of the natural world should be considered. Even if scientific experimentation or discovery suggests a supernatural cause, methodological naturalism requires that cause to be ignored or dismissed. You could practice methodological naturalism while not adopting full scientism. If you did so, you would be open to all truth except when you practice scientific research and thereby limit all your reasoning to physical explanations.

Scientific naturalism summons a greater leap of faith than methodological naturalism and is a worldview rather than a process. While methodological naturalism ignores—as a matter of professional process—any non-material cause or concept, scientific naturalism confidently declares that no such causes or concepts exist at all, anywhere, for all time. Scientific naturalism is the bold claim that everything is, was, and will be explained and contained by physical, material, natural causes. The spiritual, supernatural, or immaterial is impossible, invalid, non-existent under the worldview of scientific naturalism. In this way, scientific naturalism is closely related to scientism.

In contrast, methodological naturalism is merely a professional practice limited to scientific investigations, leaving the door open for religious personal belief. For these reasons, a theist scientist could practice methodological naturalism in the laboratory and still worship God in church. Yet a scientific naturalist denies God entirely, not just during experiments, and so does one who believes in scientism.

In summary, methodological naturalism denies the immaterial during any scientific experimentation, as a matter of practice. Scientific naturalism denies that anything immaterial exists at all, during experiments or not, believing instead that reality is limited to the physical, material world. Scientism is the most extreme position, believing that no truth exists except that which can be proven by scientific experimentation. Therefore, scientific naturalism is a belief about what exists in reality (only material, physical, natural things), while scientism is a belief about how to validate truth (only science can confirm truth).

Does scientism conflict with the common experience of humanity?

Yes. Scientism—limiting all truth to science—conflicts with the common experience of humanity that involves so much that cannot be proven by the scientific method. Science cannot prove the existence of minds, including yours or the mind of anyone you know. Science cannot prove your understanding of time and history, meaning that the scientific method cannot disprove that you were created one second ago, with false memories of the past and a stomach filled with the breakfast you remember from this morning.

Turning to experiential or emotional topics, science cannot prove that you love your husband, wife, children, or parents—or that they love you! Nor can science prove or validate the experiences that shape you, nor the individualized, personal, conscious, and freely acting essence of your being. Limited to science, you are a sack of molecules reacting to natural forces, no more than a glorified animal, and your deepest human relationships are simply artifacts of chemical, hormonal, glandular secretions during your struggle to survive. Therefore, adopting the limited worldview of scientism—excluding any truth not confirmed by the scientific method—is an impractical, myopic, and ridiculous stance once you consider the full ramifications.

Is scientism a self-refuting belief system?

Yes. Scientism is a self-refuting worldview because the validity of science itself cannot be proven or confirmed by the scientific method. Therefore, relying only upon science as a source of truth is self-defeating. Science requires many beliefs not proven by the scientific method.

Why is it impossible to use science (or the scientific method) to prove the validity or truth of science (or the scientific worldview)?

Paradoxically, scientism—relying only upon science for all truth—renders the scientific method impossible, incoherent, or sterile. The truth and reliability of both logic and mathematics are foundational requirements or preconditions for the practice of science. Yet science cannot prove logic, nor can science prove mathematics. Instead, the scientific method and scientific exploration both rely on the assumption that logic and mathematics correspond to physical reality and can be trusted. If you try to "prove" logic and mathematics using science, it is impossible because you would be arguing in a circle.

What are some non-scientific beliefs that everyone accepts, including those confused by scientism?

Even those suffering under the illogical delusion of scientism must accept beliefs that are not provable by the scientific method yet nonetheless required to practice the scientific method. Science stands on non-scientific beliefs like the existence of your mind and other minds, the reliability of mathematics, the validity of formal logic, the expectation of pattern, order, and regularity in nature, and the intelligibility of the cosmos. Therefore, believing in scientism paradoxically excludes the foundational beliefs of science itself.

Is the phrase "Trust the science" misleading or deceptive?

Yes. The phrase "Trust the science" is misleading and deceptive and therefore untrustworthy itself. Science is the exploration of knowledge, truth, cause, and effect by systematic efforts. The scientific method is an ongoing process. Science is not defined by consensus or "trust," but rather by continuous search, challenge, and refinement of knowledge. Truth exists beyond scientific confirmation.

Does science or scientific data translate into definite political solutions?

No, definitely not. The scientific method can discover knowledge about what "is," but not about what "should be." The classic distinction between "is" and "ought" (the old English version of "should be") is best outlined by the Scottish philosopher David Hume.

What did David Hume explain about the is-ought distinction?

Scottish philosopher David Hume (1711 - 1776) thoroughly examined the logical fallacy of moving from a truth claim about the world we observe (what "is") to a moral claim about how we should live our lives or govern our society (what "ought to be"). Hume warns against committing the logical fallacy of jumping from statements of fact to statements of value, also known as the is-ought problem.

Using scientific descriptions to argue for political positions often devolves into flagrant violation of Hume's is-ought distinction. Be careful to examine the political beliefs (of how things should be) smuggled into scientific descriptions (of the way things are). The very same scientific findings can support vastly different political solutions. Rather than hiding political beliefs behind scientific findings, carefully separate "what is" (scientific description of the natural world) versus "what should be" (political prescription for human action).

3 Science Opposed to Faith or Science Because of Faith?

Despite the militant atheism of some scientists today, science owes its origin to religion. Specifically, the Scientific Revolution emerged from the Christian theological belief in an ordered cosmos created and designed by God, and accessible and intelligible to mankind, who is created in the image of God and can therefore better understand and appreciate God by studying His handiwork in nature. This Christian worldview is undeniable and explicit within the pioneering minds of the Scientific Revolution, and it remains as a foundation for the continuation of modern science today. Many of the best scientific minds are and have historically been theists and devout Christians, rather than atheists and materialists.

What is atheism?

Atheism is the belief that God does not exist, and that there are no supernatural or divine entities.

Is atheism a belief? Is atheism a claim to truth? Must atheism prove its claim?

Yes. Atheism is most definitely an affirmative belief and a claim to truth. Therefore, like any other claim, the burden is on atheists to prove or defend the position.

Is atheism a "neutral" position?

No. Yet some atheists try to wriggle away from responsibility by falsely claiming that atheism is a "neutral" position. Just like belief in God (or theism), atheism (denial of the existence of God) is a positive claim that can be evaluated, justified, defended, or refuted. To be an atheist is to take a definite stance in the debate about God.

What is the "neutral" stance in the debate of God's existence?

The "neutral" position in the debate of God's existence is called "agnosticism" (not atheism, which takes a firm stance that God does not exist). If you are agnostic, you have no position or truth claim. The agnostic remains undecided and open to information and arguments about theism and atheism. Some agnostics may believe it is impossible to find the truth about God.

Is it true that the top scientists, academics, and intellectuals are all atheists? Are many of the most accomplished minds Christian or theists or religious? What is the breakdown of religiosity among Nobel Prize winners in the 20th century?

Many of the most intelligent and accomplished people believe in God and practice religion. It is a false narrative that all the top minds are atheists, or that no serious intellectual believes in God. Throughout all human history, many of the most accomplished scientists, academics, and intellectuals have been Christian, and this trend continues in our modern world.

Baruch Shalev (born 1936) studied the religious beliefs of Nobel Prize winners from the year 1901 to 2000, spanning the 20th century, in a 2005 book titled <u>100 Years of Nobel Prizes</u>. According to Shalev, more than 65 percent of the winners were Christian, more than 20 percent were Jewish, and less than 1 percent were Muslim. Notably, less than 11 percent of Nobel Prize winners were atheist or agnostic. Consider also that atheism, or atheistic materialism, is a relatively recent worldview from the last few centuries. For thousands of years of human history, the most intelligent and accomplished intellectuals were religious, and atheism was a rare, deviant, and shocking position.

When did the Scientific Revolution occur? Where did the Scientific Revolution occur?

The Scientific Revolution occurred during the 16th and 17th centuries in Europe, within the latter half of the Renaissance period. European Christendom produced the people, resources, worldview, and motivation to investigate a rational and ordered cosmos created by a God who designed consistent laws and predictable principles.

What marks the beginning of the Scientific Revolution? Who is Nicolaus Copernicus?

Most historians consider the publication of On the Revolutions of the Heavenly Spheres by astronomer extraordinaire Nicolaus Copernicus in 1543 to mark the beginning of the Scientific Revolution. Copernicus advanced the heliocentric model of the universe, whereby the sun is at the center of the galaxy, rather than the earth.

Did Nicolaus Copernicus believe in God? Was Copernicus a Christian? Was Copernicus involved in organized religion?

Yes. Nicolaus Copernicus (1473 - 1543) believed in God as the creator of the universe. Copernicus was a devout Christian who held a formal office within the Roman Catholic Church, exercising administrative and financial responsibility. Copernicus believed his heliocentric model did not contradict the Christian Bible, and he dedicated his seminal treatise to the Roman Catholic Pope Paul III.

After Copernicus, who were five of the towering figures of the Scientific Revolution? What were the summarized contributions of each of these godfathers of the scientific worldview?

Galileo Galilei (1564 - 1642) discovered the rocky, craterous terrain of the earth's moon, as well as multiple moons orbiting the planet Jupiter. By using a sophisticated telescope that he fashioned himself, Galileo also spotted countless distant stars, and observed the phases of the planet Venus, leading to his correct theory that the planets revolve around the sun.

Johannes Kepler (1571 - 1630) discovered the three laws of planetary motion and confirmed that planets revolve in elliptical, oblong courses rather than perfect circles. By tracking that planetary motion differs based on distance from the sun, Kepler paved the way for Isaac Newton's theory of gravity.

Robert Boyle (1627 - 1691) is the "father of modern chemistry" and also refined the scientific method of experimentation. Advancing the "corpuscular" theory of matter, Boyle introduced a particle and atom-based understanding of physics that corrected earlier concepts. Most famously, Boyle's law is the discovery of the inverse relationship between volume and pressure for gaseous substances.

Blaise Pascal (1623 - 1662) pioneered many concepts in geometry, probability, mathematics, economics, and social sciences. Pascal also invented the syringe, hydraulic press, and an early version of a calculator.

Isaac Newton (1643 - 1727) is responsible for classical physics because he discovered the theory of gravity itself and elucidated the three laws of motion and the law of universal gravitation. Not only did Newton advance the field of optics, co-develop calculus, and invent a reflecting telescope, his equations and laws of motion and gravity remain so influential that classical physics is referred to as "Newtonian physics."

What did Copernicus, Galileo, Johannes Kepler, Robert Boyle, Blaise Pascal, and Isaac Newton have in common, in addition to being the godfathers of the scientific worldview?

Copernicus, Galileo, Johannes Kepler, Robert Boyle, Blaise Pascal, and Isaac Newton all believed in God as the Creator of an ordered, rational universe governed by God-given laws that could be understood by mankind. For each of these titans of science, the motivation and confidence for scientific exploration came from their belief in God, and their desire to understand the handiwork and rationality of God in ordering the universe.

Why did Copernicus, Galileo, Johannes Kepler, Robert Boyle, Blaise Pascal, and Isaac Newton believe mankind could understand the universe and its laws as created by God?

Copernicus, Galileo, Johannes Kepler, Robert Boyle, Blaise Pascal, and Isaac Newton (and countless others) all believed the human mind was created in the image of a rational God. This correspondence between the rational mind of God and the rational mind of mankind generated the scientific impulse to discover, interpret, and understand the laws of nature.

What did Galileo Galilei say about the gifts from God to humanity of sense, reason, and intellect?

Galileo Galilei wrote to the Duchess Christina of Tuscany:

"I do not feel obliged to believe that the same God who has endowed us with sense, reason, and intellect has intended us to forgo their use."

What did Johannes Kepler say about discovering the mind of God by exploring nature?

Johannes Kepler wrote in a letter to his colleague, Herwart von Hohenburg:

"God wanted us to recognize [the order in nature as established by God] by creating us after his own image so that we could share in his own thoughts."

What did Robert Boyle say about the relationship between God, nature, and scientific exploration?

Robert Boyle is known for several quotes about his God-inspired scientific curiosity, including:

"God would not have made the universe as it is unless He intended us to understand it."

"God [is] the author of the universe, and the free establisher of the laws of motion."

"From a knowledge of His work, we shall know Him."

How did Blaise Pascal use applied mathematics and probability to estimate the benefits and risks of faith in God?

"Pascal's wager" refers to Blaise Pascal's observation that faith in God offers eternal, glorious reward at the possible risk of being incorrect upon death, after leading a just and moral life. Yet the atheist gains nothing if proven correct but forgoes eternal life if proven wrong. In the words of Pascal in his famous book, Pensées (French for "thoughts expressed in writing"):

"Belief is a wise wager. Granted that faith cannot be proved, what harm will come to you if you gamble on its truth and it proves false? If you gain, you gain all; if you lose, you lose nothing. Wager, then, without hesitation, that He exists."

How did Isaac Newton describe the solar system that he studied to produce the theory of gravitation and the laws of planetary motion, revolutionizing the field of physics and astronomy?

Isaac Newton was inspired by confidence that a rational God created an ordered universe and that the glory of creation is God's unmistakable signature:

"This most beautiful system of the sun, planets and comets, could only proceed from the counsel and dominion of an intelligent and powerful Being."

"Gravity explains the motions of the planets, but it cannot explain who sets the planets in motion."

"He who thinks half-heartedly will not believe in God; but he who really thinks has to believe in God."

Did the Scientific Revolution emerge within an intellectual environment of Christian theism studying a rational God who created an ordered cosmos? Was Christian theology, including the creation of mankind in the image of God, a causative factor in early scientific exploration, not simply a coincidence of timing?

Yes. The Scientific Revolution was born out of the Christian understanding of a rational God who created the universe according to ordered laws. The earliest pioneers of the scientific method thought they could understand the laws of nature because God infused order into the world, and the rational intellect of mankind could glimpse into the rational mind of God by exploring the cosmos. Therefore, Christian doctrines of (1) Creation of an ordered universe by a rational God and (2) mankind created in the image of God were directly linked to the earliest scientific impulses and endeavors.

What prerequisites to the birth of science did the physicist Peter Hodgson identify?

The British physicist Peter Hodgson (1928 - 2008) explained in "The Christian Origin of Science" (1997):

"If we think about what is needed for the viable birth of science, we see first of all that it needs a fairly well-developed society, so that some of its members can spend most of their time just thinking about the world, without the constant preoccupation of finding the next meal. It needs some simple technology, so that the apparatus required for experiments can be constructed. There must also be a system of writing, so that the results can be recorded and sent to other scientists, and a mathematical notation for the numerical results of measurements. These may be called the material necessities of science."

Therefore, Hodgson identified an organized society with the luxury of living beyond mere subsistence and food-gathering, basic technology, literacy of the written word, and mathematical notation as the material prerequisites to the birth of science.

Civilizations and societies before 16th- and 17th-century Europe had technology, writing, and mathematics, so why did science fail to originate in those earlier environments?

The missing ingredient was belief in the Christian God, and the expectation of an ordered, rationally created universe that is intelligible to the mind of humanity as endowed in the image of God.

The ancient Indian, Chinese, Greek, and Roman civilizations, empires, and societies had technology, writing, and mathematics, but no conception of a monotheistic, personal, creative God who endowed mankind with the divine image and rationality of God. Without this theological background, earlier civilizations lacked the impulse, justification, and reasoning to explore the laws of the universe as the work of the mind of God. These ancient civilizations had the material goods, but not the intellectual and theological catalyst.

How did the British mathematician and philosopher Alfred North Whitehead describe the theological catalyst for the birth of science?

British mathematician and philosopher Alfred North Whitehead (1861 - 1947) argued in his book, Science and the Modern World (1925), that the scientific impulse draws upon a belief in the order of nature, based upon the rationality of God:

"There can be no living science unless there is a widespread instinctive conviction in the existence of an Order of Things. And, in particular, of an Order of Nature."

"It is this instinctive conviction, vividly poised before the imagination, which is the motive power of research: —that there is a secret, a secret which can be unveiled. How has this conviction been so vividly implanted on the European mind? When we compare this tone of thought in Europe with the attitude of other civilisations when left to themselves, there seems but one source for its origin. It must come from the medieval insistence on the rationality of God..."

How did C.S. Lewis describe the scientific impulse to discover the laws of nature?

British scholar, author, and theologian C.S. Lewis (1898 - 1963) wrote in one of his many books, Miracles (1947):

"Men became scientific because they expected Law in Nature, and they expected Law in Nature because they believed in a Legislator."

How did the American philosopher Holmes Rolston III explain the monotheistic intellectual background of the pioneers of scientific thought?

American philosopher Holmes Rolston III (1932 - 2025) stated in Science and Religion (1987):

"It was monotheism that launched the coming of physical science, for it premised an intelligible world, sacred but disenchanted, a world with a blueprint, which was therefore open to the searches of the scientists. The great pioneers in physics—Newton, Galileo, Kepler, Copernicus—devoutly believed themselves called to find evidences of God in the physical world."

What is the historical connection between religion and academia? What is the origin and etymology of the term "university" used for institutes of higher education? Were the first universities distinctly Christian institutions?

Although much of modern academia seems hostile to religious belief, ironically, higher education is a product of Christianity. The title for institutes of higher education, "university," has a distinctly Christian origin story. The word "university" comes from the Latin words, "uni veritas," meaning "one truth" in English.

The earliest universities in medieval Europe trained students in the specialized canon law of the Roman Catholic Church, Christian theology, and the humanities. In Great Britain, both Oxford (founded in the 11th century) and Cambridge (founded 1209) have Christian origins. John Harvard was a former Cambridge student who emigrated to America to found Harvard University in 1636. American universities (or colleges) with a Christian origin also include Yale, Brown, Princeton, William & Mary, Columbia, Darthmouth, Notre Dame, Pepperdine, Georgetown, Baylor, Colgate, Chicago, Duke, Syracuse, Loyola, Liberty, and Wheaton, among many others.

What scientific or cultural ideas led to a distrust of theology and religion? When did the friction between religion and science begin?

The three most obvious examples of cultural ideas leading to a decline of religiosity come from biologist Charles Darwin, political philosopher Karl Marx, and psychologist Sigmund Freud. All three intellectual currents arrived in the late 19th and early 20th centuries.

How did the ideas of Darwin, Marx, and Freud erode religion and belief in God?

Darwin, Marx, and Freud offered material explanations for life (including human origins), human societies, and human psychology that seemed to eliminate the need for God and therefore religion. Others have summarized the trio as follows: Freud told us who we are and how to justify our sins, Darwin told us where we came from, and Marx told us where we are going.

How did the ideas of Charles Darwin erode religion and belief in God?

In 1859, Charles Darwin (1809 - 1882) published On the Origin of Species, which provided an evolutionary explanation for animal life and human beings as derived by natural selection acting on random variation. Darwinian evolution argues that what seems like design by God is an illusion instead produced by material forces. Notably, Darwinian evolution characterizes mankind as nothing more than an evolved animal, sharing an ancestor with apes, rather than an intentional creation in the image of God.

How did the ideas of Karl Marx erode religion and the belief in God?

Karl Marx (1818 - 1883) published his famous pamphlet, "The Communist Manifesto," in 1848. The seminal treatise of Marx, Das Kapital ("The Capital" as in the word, "capitalism") was published in three volumes: the first in 1867, and the second and third after Marx died in 1885 and 1894, thanks to his collaborator, Friedrich Engels (1820 - 1895).

Marxist theory offers a materialist conception of history, and Engels coined the term "historical materialism." Following Marxism and historical materialism, human history is driven by material, technological changes that spur social progress based on the social classes that struggle over unequal access to economic capital. Marx famously quipped that "religion…is the opiate of the people," diminishing belief in God as a foolish illusion of the ignorant masses.

How did the ideas of Sigmund Freud erode religion and belief in God?

Psychologist Sigmund Freud (1856 - 1939), confirmed atheist, introduced his form of psychoanalysis, focusing on unconscious psychological drives and impulses mostly stemming from formative childhood events and repressed sexual urges. Therefore, Freud explained the human condition based on material forces of the environment and body, not the spirit, soul, or image of God endowed to mankind. Morality, sin, and salvation were put aside in favor of materialist, biological, psychological explanations for human choices, feelings, desires, and relationships.

Historically, how did science and faith in God become distorted to be seen as in tension rather than in harmony? Which two books were highly influential in revising history to reflect a constant conflict or warfare between religion and science?

Two books published in the late 19th century diminished or overlooked the essential roles played by Christian belief and theology in the beginning of the Scientific Revolution. Instead, these books recast the historical relationship between religion and science as being enemies in perpetual "conflict" and even "warfare" according to the book titles.

John William Draper (1811- 1882) published History of the Conflict Between Religion and Science in 1874. Andrew Dickson White (1832 - 1918) published A History of the Warfare of Science with Theology in Christendom in 1896. Published amid the cultural upheaval following Charles Darwin's treatise on the theory of evolution published in 1859, both books inaccurately portrayed science as the adversary, rather than the result, of the Christian worldview. Draper wrote:

"...Christianity and Science are recognized by their respective adherents as being absolutely incompatible; they cannot exist together; one must yield to the other; mankind must make its choice—it cannot have both."

4 The Scientific Mind and its Dependence on Philosophy, Theology, and Immaterial Concepts

Reducing all science to materialism forgets that scientific inquiry comes from immaterial minds, not unconscious matter. Further, science itself is not possible without certain philosophical and theological beliefs. Therefore, it is false and misleading to claim that science is independent from philosophy or theology. Rather, science requires both philosophy and theology, and science requires recognition of immaterial concepts.

Did matter or mind first open the door to scientific exploration?

Immaterial minds, not physical matter, opened the door to scientific exploration. It was human consciousness, curiosity, rational contemplation, and faith in an ordered universe with regular laws that persist across time and space that bridged the gap between the natural world and our growing knowledge thereof. All scientific knowledge is the description and distillation of the natural world through observation and experiment. Yet this knowledge is not part of the material world and must be understood and shared by immaterial minds.

Is scientific knowledge material or immaterial? Is matter or mind behind the origin of science?

All knowledge, even scientific knowledge, is ultimately immaterial. We search, grasp, contemplate, and synthesize this knowledge with our immaterial minds. The physical media of shared science—books, articles, recordings, videos, data troves—exist only to connect to the maximum number of immaterial minds across space and time. We share scientific knowledge through material media only because mass telepathy is not available, but the fact remains that science is an immaterial concept that studies a material world. Mind, not matter, is the origin, vehicle, and destination of science.

Why has some portion of the scientific community turned its back on the immaterial origins of science? Has the materialist worldview produced stunning success?

While the immaterial mind is the origin of science and excluding non-material causation from science limits progress, focusing exclusively on materialist models has been immensely successful. Nobody can deny that purely mechanical and materialist discoveries have made impressive strides in medicine, engineering, and technology.

The materialist worldview has achieved stunning success in life expectancy, hygiene, food production, astronomical exploration, telecommunication, transportation, energy, and much more. Yet the mind and careful contemplation of immaterial concepts sparked the embers of the roaring fire we now enjoy. The success of materialist ventures in some areas must not close the door on the study of primary and ultimate questions: philosophy and theology. In fact, science is impossible without philosophy and theology.

Why is science impossible without philosophy? How is philosophy a required precondition of scientific exploration?

Conducting science requires several philosophical commitments. First, you must believe in your ongoing, personal, conscious identity, and in the validity of your perception of and interaction with the outside world, including with other people and objects. Only through philosophy of knowledge (epistemology) can you trust your sight, hearing, thoughts, or anything outside of your own mind. The person who sets up an experiment is ideally the same person who records the results and then evaluates, analyzes, and derives conclusions. Despite the need for continuity of identity, you cannot prove your continuing, conscious identity—and the validity of your perceptions across time—using the scientific method. Instead, you accept these truths through your philosophy of mind, metaphysics, and epistemology.

Second, the idea that nature exhibits regular, observable laws is a philosophical position. We take this philosophy for granted, but philosophy it is. We cannot prove that tomorrow will resemble today, or that gravity or electromagnetism will persist uniformly, but we expect it to be so by philosophical belief, supported but never proven by scientific observation. Tomorrow is always another day.

If we instead believe that the cosmos is so chaotic that fundamental laws and forces change at any moment, scientific experimentation is impossible. Fixed, controlled, predictable conditions are at the heart of properly constrained experiments. Our philosophy of a rational, ordered, predictable, and observable cosmos does not transform into necessity simply because its familiarity made us forget about its contingency.

When were scientists first called scientists?

Nobody performing what we recognize as scientific research was called a scientist until at least the year 1833, when William Whewell first used that term.

Who was William Whewell?

William Whewell (1794 - 1866) was a polymath and philosopher of science who coined the term "scientist" in the year 1833 and "physicist" in 1840.

What were scientists called before the term scientist was ever used?

Philosophers. Before the term scientist emerged in the year 1833, those performing what we recognize as scientific research were called "experimental philosophers," "natural philosophers," or "mechanical philosophers." These earlier names reveal that science is a subdivision within the wider philosophic study of knowledge, reality, and existence.

Why is science impossible without theology? How is theology still practiced within science despite the atheistic materialism of some scientists?

Theology—the study and knowledge of God—is inherently linked to science because it gave birth to science and pervades the scientific process. The origin of science as we know it comes from the theological position that our cosmos is ordered, rational, and intelligible. The cosmos is ordered because it is designed and created by a God who endowed mankind with the same rationality that makes the laws of nature intelligible, and therefore fruitful to study, test, and analyze.

The atheistic materialism of scientists is also a theological position: God does not exist, and causative explanations that point to God or to the supernatural should be ignored. Disbelief in God is a theological belief, and strict exclusion of the supernatural is a method that stands on a theological position (called methodological naturalism).

In contrast to atheistic materialism, the founders of science believed in God and understood that supernatural causation and creation were fundamental to understanding the natural world. The fact that since the 19th and 20th centuries, many modern scientists changed their theological stance to deny God cannot erase the existence of theology within science.

How would the professional community of scientists behave differently if theology played no role in science? What would genuine theological neutrality look like?

Atheism is a theological belief. If theology played no role in science, scientific evidence or inferences of God or the supernatural would not result in career suicide for the scientist. Instead, explanations of supernatural or divine causation would be evaluated objectively, without penalty or insult, along with any other hypothesis.

Instead, a dismissive strain of theology within science threatens persecution, insults, and career-ending backlash heaped upon scientists who follow evidence, data, and observations to anything remotely related to God or the supernatural. Agnosticism or indecision is neutral; belief (theism) or disbelief (atheism) is a firm theological position. Any explanation, whether physical or immaterial, should stand on its merits rather than being prejudged by theological prejudice.

5 How Did We Get to This Sterile, Narrow Worldview? The Expulsion of Purpose from Scientific Thinking

The atheistic materialist worldview ignores or denies purpose and instead focuses only on physical mechanisms. Yet our everyday experience as well as the entire universe (and therefore, all scientific research) is meaningless and impossible to explain without referencing purpose. Limiting scientific exploration to only material and physical causes ignores the most enlightening questions of purpose, design, and meaning. Perception requires a point of view, experiments are designed for purpose, and therefore science will never achieve its unrealistic promise of objectivity. Despite modern efforts, purpose cannot be avoided in the search for truth.

How is purpose necessary for everyday experience?

Without purpose, you have no context for any decision, action, or understanding. Purpose is infused in everything you recognize, but purpose is so common that it is taken for granted if not forgotten entirely.

How is purpose infused in everything you recognize?

You distinguish objects by purpose. The difference between an abstract wood sculpture and a chair is that you know you can sit on a chair. Your recognition of an identifiable object or process is inherently tied to your ability to assign a purpose.

How could you describe the process of making a cup of tea without reference to any human purpose?

Consider the preparation of a cup of tea. Rather, contemplate this process when stripped of any purpose. If you describe the process "scientifically" or "mechanically," you adopt the awkward posture of atheistic materialism.

You were not fetching a hot drink for yourself or a guest. Rather, using a heat-generating device, you excited the molecules of water trapped within a stainless-steel container. Then, you distributed the boiling-hot water into a smaller container and allowed dried leaves to partially disintegrate into the solution for 300 seconds. Finally, you dissolved polysaccharides ("sugar") and mammary gland secretions ("milk") using a metal instrument shaped into an oval at the end of a handle ("spoon").

Stripped of all context, meaning, and purpose, this description of making tea sounds like the product of a diseased mind, a robot, or an alien from another planet. On the other hand, the easiest and most straightforward description is that you prepared a cup of tea for yourself or a guest. The mechanism of process (heating, mixing, adding ingredients) is subordinate to the ultimate purpose (getting a cup of tea to drink). The "how" is not as important nor as descriptive as the "why."

Why is purpose essential, required, and unavoidable in scientific exploration?

Without purpose there is only description and data collection. But even meaningful description requires context and comparison to an existing worldview. With no purpose or context, scientists can record results but make no conclusion, hypothesis, or synthesis with existing knowledge. Without purpose, we drown in meaningless data.

Purpose is infused in everyday life and just as much in the practice of science. Every scientific experiment has a purpose no matter how carefully the researchers hide it. Objectivity in science is total nonsense. Purpose, point of view, expectations, and subjective experience drive the beginning, middle, and end of all scientific activities.

Does the scientific process focus more on "how" questions than on "why" questions?

Yes. For centuries, the scientific process has avoided the ultimate questions of "why" – which speak to purpose – in favor of intermediate questions of "how." Exploring and answering questions of "how" lead to descriptions that are helpful yet incomplete. To get the full story and comprehensive context, you must eventually consider purpose.

When was purpose expelled from scientific thinking?

In the 17th century, during the rise of the mechanistic and materialist worldview, purpose was expelled from orthodox scientific thinking. Studying purpose answers the question of "why." Studying mechanical causes and effects instead answers the question of "how." Mainstream science prioritized mechanistic questions of "how" after significant intellectual and scholarly debate during the 17th century.

Why was purpose expelled from scientific thinking?

The leading scientists of the 17th century argued that scientific inquiry, observation, experimentation, and analysis should focus on the materialist, mechanistic "how" rather than the "why" (or purpose) of the physical universe and living organisms.

Drawing attention to the purpose for physical behavior, whether of animals or material objects, seemed to contradict or transcend the dogmatic scientific worldview that permits only unconscious physical matter. Ideas about the "life force" or vitalism or soul were rejected outright. Insights about a creative force or God or higher power or designer were similarly rejected as "unscientific" and therefore unworthy of the attention of scientists.

What is another term for purpose in discussing causation of the natural world?

Another term for purpose is "final cause" or "ultimate cause." While mechanical explanations focus on the physical push and pull of concrete, tangible objects, explanations of purpose, final cause, and ultimate cause go beyond mere physical impetus.

What is teleology?

Teleology is the study of purpose and design—precisely the field of inquiry that modern science ignores, ridicules, and suppresses.

What is one criticism of relying only upon final causes or purpose-driven scientific explanations?

One persuasive critique of final causes or purpose-driven explanations within scientific thinking is the danger of falling into circular reasoning that provides no further insight. Ancient and medieval philosophers and scientists explained phenomena by the "nature" of the entity or event. If asked to explain why bread is nourishing to the human body, the answer was that bread had a nutritive or nutritious "nature." Therefore, the inherent quality or nature of bread was to provide nutrition. If asked to explain why fire transmitted heat to nearby objects, the answer was that fire has the property or "nature" of heat and warmth.

The problem with these circular explanations is that they provide no deeper insight into the process or quality of the phenomenon. Stating that nutritious food has a nutritious nature, or that hot flames have a hot nature, is nothing more than a restatement of the original inquiry. Partly for these reasons, many scientists, philosophers, and intellectuals in the 17th century made a concerted effort to reject the exploration of final cause or purpose.

What are the four classifications of causation according to Aristotle?

The ancient Greek philosopher Aristotle (384 - 322 BC) classified causation into four categories that correspond to purpose, form, matter, and energy. Aristotle identified these four types of causes to explain why things change, move, or come into being. The four types of causation are efficient cause (energetic cause or energetic impulse), material cause (matter, substance), formal cause (form or shape), and final cause (purpose).

Did Aristotle use the word "cause" differently from how that word is used in modern English?

Yes. The common, modern definition of "cause" may confuse these concepts introduced by Aristotle. A better word for what Aristotle meant might be "explanation," or the four classifications of explanation.

Of course, Aristotle wrote in the ancient Greek language. The English word "cause" is commonly translated for the actual word used by Aristotle, which is "aitia." The better translation for "aitia" may be "explanation," at least in this context.

What is the efficient (or energetic) cause within Aristotle's four types of causation?

The efficient cause brings about change usually through physical forces acting upon the object in question. Mechanical explanations about the push and pull between and among tangible objects relate to efficient cause. You can also consider the efficient cause to be the energetic cause because something expends energy to make changes.

For example, the efficient cause of a wooden rocking chair is a carpenter who fashions that chair. The efficient cause is sometimes called the "agent" cause to highlight the source of the action that changes or creates the object.

What is the material cause within Aristotle's four types of causation?

The material cause concerns the substance or matter that determines the change or movement. For example, the material cause of a wooden rocking chair is, of course, wood derived from trees. Material cause recognizes that the substance and structure of different materials will react in different ways to the same stimulus. Dense and durable materials like steel or stone will not break under significant pressure, but delicate materials like eggshells, glass, and thin paper will crack, shatter, and tear apart.

What is the formal cause within Aristotle's four types of causation?

The formal cause concerns the "form," shape, arrangement, or visual appearance of the object in question. For example, the formal cause (shape) of a wooden rocking chair corresponds to the recognizable design and appearance: a flat seat, curved rocking legs, armrests, and other key structural elements.

What is the final cause within Aristotle's four types of causation?

The final cause deals with the purpose of the object or event. Explanations of purpose and Aristotelian final causes have been largely ignored by modern science.

How did Saint Thomas Aquinas refine our understanding of Aristotle's four types of causation?

Theologian and philosopher Saint Thomas Aquinas (1225 - 1274) believed that the final cause or purpose is the ultimate or most important cause of all. To Aquinas, matter is further refined by form, and form is provided by the efficient causal agent. But the final cause—the purpose or design or end-goal—is the primary source of all other causation. Therefore, the final cause or purpose is paramount to understanding all other forms of causation.

What revolutionary book spurred the expulsion of purpose from scientific thinking?

In the 17th century, Sir Francis Bacon (1561 - 1626) published The Advancement of Learning (1605), which argued that scientific exploration should focus on efficient (energetic), material, and formal causes, and should reject or ignore final causes related to purpose. Bacon explained that scientific inquiry should examine the tangible, material aspects of the natural world, and that final causes improperly distracted the scientist by speculation into abstract, intangible purposes.

What is the inherent weakness in the elimination of final causes (purpose) from scientific thinking?

Most explanations require purpose; purpose is inescapable. Observation of nature itself, especially that of living organisms, invites speculation about the final cause or purpose of various attributes or behaviors. For example, many animal behaviors and body parts seem impossible to explain or understand without referring to purpose (final cause).

The purpose or final cause of a spider spinning a web is, of course, to catch flies for the spider to eat. The final cause of a bird building a nest is to construct a home, perhaps to incubate eggs in a safe place. The purpose of gills is to breathe underwater, and that of fins, to swim through water.

It strains credulity to notice so many clues and signposts pointing to goals, purposes, and final causes, but to dismiss any consideration thereof due to preconceived notions of what is supposedly (un)scientific. You may instead embrace the idea that purpose is evident and obvious within nature, within the universe, and within every living organism.

How do nature documentaries and other educational presentations about the natural world reveal the importance of final causes, and therefore, purpose?

When you watch a nature documentary, notice that the narrator is forced to provide final-cause explanations about purpose and end-goals for all living organisms. Challenge yourself to find any nature documentary that does not refer to end-goals or purposes of animal behavior and physiology. If you do find such a nature documentary, stripped of any description of purpose, you will have found the most boring and incoherent film available. We cannot understand nature—whether for investigation or entertainment—without considering purpose.

PART 2, MATTER AND COSMOS

6 Introduction to the Immaterial Nature of Our Universe

For thousands of years, the best minds wrestled over the finitude versus eternity of the universe. Did the universe come into being, or is it eternal?

If our cosmos has always existed, atheists and materialists are relieved, because there is no need to explain the creation of the cosmos if there is no beginning. In the mid-20th century, the scientific community confirmed that our cosmos had a distinct beginning—precisely what Christianity taught from the first century, and what the Israelites of the Old Testament taught for centuries earlier. The atheist astronomer, Fred Hoyle, derided the notion of this spectacular appearance—from nothing—of all space, time, matter, and energy as "the Big Bang." The name stuck. Hoyle was right to become defensive, because his atheist position had the rug pulled out from underneath it, and science itself was responsible.

Not only did science confirm that the cosmos had a beginning, just as the book of Genesis in the Bible and Christian theology describes, this explosive event posed mind-boggling challenges to the atheistic materialist worldview. All matter, all energy, all time, and all space emerged in a single instant—from nothing—and expanded over billions of years into the spectacular order of galaxies, stars, planets, and lives we experience today. Strict materialism cannot explain our universe coming into being, because the Big Bang confirms that matter itself first appeared some 13.8 billion years ago. Before the Big Bang, there was no matter to provide any material (matter-based) explanation for the creation of the cosmos. From the very beginning of everything studied by science in the cosmos, atheistic materialism is ruled out as a viable worldview.

Further eroding the credibility of the strict materialist worldview is the stunning revelations of the infinitesimally fine-tuned conditions of our cosmos. Whether you examine gravity, electromagnetism, the strong and weak nuclear forces, the cosmological constant of the expansion of the universe, or scores of other constants, laws, and delicately balanced equations of physics, you discover mind-bogglingly precise fine-tuning. These values, constants, equations, and relationships could have had trillions of other combinations, but each aligns with shocking precision within a hair's breadth that allows for our cosmos to exist with galaxies, stars, planets, and the elements required for life. Life itself cannot exist except under the fine-tuned conditions that were established from the initial explosion of all matter, time, energy, and space "in the beginning."

Fine-tuning has confounded and challenged atheist physicists and philosophers so much that some launch into fantastical ideas about multiple or infinite alternate realities—the same Multiverse fantasy encountered in comic books and movies. The brute, mathematical and physical facts of our incredibly balanced, fine-tuned universe pose unquestionable odds against everything falling into place by chance. The probabilities and numbers are so badly stacked against chance—and therefore the atheistic materialist worldview—that desperate holdouts try to reimagine reality itself to avoid this firm conclusion.

Dreaming up the evidence-free Multiverse, skeptics fine-tune an infinite number of alternate, never-detected realities in a misguided and laughable effort to monkey with statistics and probability. If there are infinite other universes not yet detected, with an infinite range of other values, constants, equations, and forces, then our incredibly unlikely cosmos would seem less special, less rare, less finely tuned. How sad that some will abandon a rational, accommodating universe in favor of fantasies and fabrications of alternate realities within the Multiverse.

Meanwhile, back on planet earth, within our cosmos and reality, all the signs of fine-tuning point to a fine tuner. The abrupt appearance of a Singularity of all matter, all energy, all space, and all time requires an immaterial, infinitely powerful, timeless, transcendent, personal creator to manifest and pull the cosmic rabbit of all reality from nothingness, not even from a hat. This magician, this Creator, is beyond mankind but reveals a rational quality of mind that mankind can perceive, even if "through a mirror, darkly." The necessary qualities of this Creator who summoned all reality and left a signature for mankind to interpret points directly to the basic definition of what everyone considers to be God.

How profound it is that the ordered, rational cosmos, showing unmistakable elements of design and intelligence, is itself intelligible and understandable to mankind, created "in the image of God." This principle of correspondence between the mind of man and the mind of God, discoverable through exploration of nature, is the intellectual bridge that led to science. No matter how many atheistic materialists wish to burn that bridge, the arc of history rewinds back to the godfathers of science who believed that God is our Father.

More unwelcome news for the atheists and materialists: Matter itself is not as solid or reliable as it seems. Imagine placing all your faith (and it is faith) in a worldview that relies exclusively on material objects and matter, excluding all other causes. Scientific discoveries of the 20th century were a disaster for the atheistic materialist worldview. First, the Big Bang confirms the sudden appearance of everything you thought explains our reality: all matter, energy, time, and space burst into being from nothingness. Second, quantum mechanics investigates the nature of matter itself and destroys the foundation of materialism.

The tiniest building blocks of matter are in constant change, constant motion, and cannot be predicted or plotted. Matter itself at the smallest quantum level is indeterminate and can only be described, awkwardly, in a probabilistic wave function of possibility. As Werner Heisenberg noticed, Aristotle's idea of potentiality is confirmed centuries later by the strange, surreal, and uncanny behavior of quantum particles that exist somewhere and nowhere at the same time. Worse yet, at the greatest level of material detail, particulate matter can only be described by mathematical equations and has no fixed points or predictable, linear paths.

Drill down enough with the strongest microscopes and you discover a world that is intelligible only through the abstract language of mathematics, representing the ultimate victory of the idealists over the materialists. How strange and uncomfortable for the materialists that material is not concrete, not solid, not predictable, not even stable in any meaningful sense...and can only be understood by the same immaterial mind that strict materialists deny. At the scientific core of all matter we rediscover the immaterial mind, the only entity who can understand abstract, intangible, immaterial concepts written in the mathematical language of quantum physics. Mind, matter, consciousness, and cosmos are connected just as much by science as they are by religion, philosophy, and the common experience of humanity.

7 The Beginning of Everything: The Big Bang

Science confirms the cosmos emerged suddenly from nothing, settling the ancient debate of whether the cosmos was eternal or finite. At the moment of the Big Bang, all matter, all energy, all time, and all space came into being from a mysterious state of nothingness. Atheistic materialism and scientific naturalism – limited to material causes – have no ability to explain the beginning of the entire cosmos. No material cause can explain the Big Bang because the Big Bang marks the beginning of all matter. When you examine the necessary properties of who or what could generate the Big Bang, the description precisely matches the common understanding of God: transcendent, immaterial, timeless, spaceless, changeless, incredibly powerful, all-knowing.

What is the scientific concept or name for the beginning of everything in the cosmos?

Scientists call this beginning event of everything in the entire cosmos the "Big Bang."

What do you mean by the beginning of "everything" in the entire cosmos? When did the Big Bang happen?

The Big Bang is the moment some 13.8 billion years ago when all matter, energy, time, and space came into being after appearing from a state of absolute nothingness. Please re-read the last sentence and consider it carefully.

What is the Big Bang? What is the Singularity?

The Big Bang refers to the earliest moment of time, in fact, when time itself started. All matter in the universe, all energy in the universe, all time, and all space came into being, suddenly, and exploded from an infinitely dense "Singularity." This explosion (Big Bang) expanded rapidly and continues to expand right now. This expansion over billions of years formed the earth, all the planets, and all material and energy (and all time and space) we observe throughout the universe.

Does the science of astronomy, cosmology, and physics confirm that before this moment of the Big Bang, neither time nor space nor matter nor energy existed? How does the Big Bang align with traditional religious beliefs?

Yes. The scientific understanding of the Big Bang is that all space, all time, all energy, and all matter came into being at a distinct point. Before this Big Bang event, there was no time, no space, no energy, and no matter.

To be clear, this is merely the scientific explanation according to physics, astronomy, and cosmology. You may notice that both materialist science and Christianity agree that (1) the universe is not eternal (2) the universe emerged in a sudden, dramatic event and (3) before the universe emerged, there was no matter, no space, no time, no energy.

Does the Big Bang leave unexplained the biggest question of all: How did the entire universe come into being...from nothing?

Yes. The Big Bang describes the events of the universe expanding *after* everything appears, but the Big Bang does not explain nor try to explain how something (really everything) came from nothing.

Who or what can possibly exist before the existence of time, space, matter, and energy?

Good question! Before the Big Bang, there was no time, space, matter, or energy. The profundity of this absence cannot be overstated. Before literally everything emerged at once, there was nothing…not anything, and nothing to describe as empty. By "nothing" we do not mean empty space or a cluster of raw, unformed material. In this context, "nothing" takes a special meaning excluding time itself, the boundaries of physical space, and all matter and energy that has ever existed. Only someone or something beyond time, space, matter, and energy could exist "before" those fundamental elements appeared.

Does the Big Bang model offer any explanation of how everything came into being from nothing?

No. The Big Bang model explains only what happened after all matter, energy, time, and space already appeared. The Big Bang model offers no explanation for how everything appeared from absolutely nothing in that special moment.

Is it possible for the atheistic materialist worldview to explain how all matter and energy in the universe came into being from nothing?

No. By definition, atheistic materialism cannot explain how all matter and energy came into being from nothing. All materialist explanations are impossible because there is no matter or energy available to do anything, or to create anything, or to cause anything, prior to the Big Bang event. Therefore, atheistic materialism fails as an explanatory worldview because it is limited to causation by matter or energy of the material world.

Before the Big Bang, there was no matter or energy, there was truly nothing at all. This absolute absence of any material exposes the severe limitations of atheistic materialism for the origin of everything in our cosmos. If there is no matter or energy, and only mechanistic, materialist explanations are valid, then the worldview of atheistic materialism is impotent and has nothing to offer about the origin of the cosmos.

Does the Big Bang model explain how time and space came into existence?

No. The Big Bang model explains only what happened immediately after both time and space emerged from a mysterious nothingness. Before the Big Bang, there was no time to mark the arrow of temporal progress, nor was there any space to contain matter or energy. The Big Bang model offers no explanation for how time and space emerged but instead takes as an initial condition the incredible inception of both time and space.

What is the glaring weakness in stopping all intellectual and scientific inquiry at the moment of the Big Bang after all time, space, matter, and energy came into being from a mysterious original state of absolute nothingness?

The Big Bang model is all about the "bang" (explosion and expansion), but the model says nothing about what "set the explosive charge" or initially provided the elements or catalyst to create the material and energy that eventually populated our known universe over billions of years. While the Big Bang model is incredibly complex, remarkably explanatory, and marvelously illuminating, the description answers only what happened right after the most astonishing phenomenon of all existence in our cosmos. Right before the Big Bang is when everything appeared from nothing.

What is the most astonishing phenomenon of all existence in our cosmos? What is the greatest scientific mystery from which all other questions arise?

Something—in fact, everything—came from nothing. The origin of the cosmos is the greatest scientific mystery from which all other questions arise, like the origin of life, and the origin of the mind and consciousness. The first, original, most important phenomenon of all physical existence is how or why the entire universe we can study, contemplate, experience, observe, and perceive came into being in the first moment. "In the beginning" appeared all matter and energy throughout the entire cosmos, as well as the original inception of both time and space.

Since atheistic materialism cannot explain the origin of the cosmos, what alternative worldview can explain the origin of the universe from nothing?

Theism, or the belief in a creative God, can explain the origin of the cosmos. While theism is the general belief in God, there are many different religions that believe in different types of gods.

What theistic religion has the strongest claim to predict and explain the origin of the cosmos?

Christianity is the theistic religion with the strongest claim to predict and explain the origin of the cosmos. The Christian Bible describes the Creation event of the cosmos—everything visible and invisible—by God, in the ancient, first book of the Old Testament, Genesis, chapter 1:

"In the beginning God created the heavens and the earth."

Why does Christianity have the strongest claim to predict the origin of the universe?

For centuries, philosophers and scientists have debated whether the universe is eternal or instead finite, beginning at a certain point. If the universe is eternal, there is no need to explain its origin, because the universe has no beginning.

For thousands of years, the first chapter of Genesis in the Old Testament of the Bible has established the claim that God created the universe. Christianity therefore holds that the cosmos is finite, not eternal.

Scientific observation and analysis in the mid-20th century confirmed the Christian worldview of a finite universe with a definite beginning. This fundamental tenet of Christianity – "Creatio ex niholo" in Latin, "Creation from nothing" in English – predated the scientific confirmation of the finitude of the universe by thousands of years. Other non-Christian, non-Abrahamic religions have creation myths, but they typically involve a deity manipulating existing matter, rather than Creation from nothing. Other religions instead predicted an eternal cosmos. Therefore, the Christian religion correctly predicted the shockingly sudden emergence of the entire universe, from nothing, and was eventually confirmed by scientific means, thousands of years later.

How did astronomer Robert Jastrow describe the cosmological confirmation of Christian Creation theology by the scientific discovery of the Big Bang?

In his 1978 book, God and the Astronomers, astronomer Robert Jastrow (1925 - 2008) poetically described the stunning confirmation by scientific cosmology of the Christian doctrine of Creation:

"For the scientist who has lived by his faith in the power of reason, the story ends like a bad dream. He has scaled the mountains of ignorance, he is about to conquer the highest peak; as he pulls himself over the final rock, he is greeted by a band of theologians who have been sitting there for centuries."

How did Nobel Laureate Arno Penzias describe his co-discovery of evidence for the Big Bang in relation to Christian theology?

Physicist and astronomist Arno Penzias (1933 - 2024) co-discovered with Robert Wilson (born 1936) the cosmic microwave background radiation that proved the Big Bang, which earned them both the Nobel Prize in 1978. Penzias described the scientific evidence for the Big Bang as coinciding precisely with the Biblical doctrine of the Creation of the cosmos by God:

"The best data we have [about the Big Bang] are exactly what I would have predicted, had I nothing to go on but the five Books of Moses, the Psalms, the Bible as a whole."

According to fundamental Christian doctrine, who created the universe and everything within the universe?

The fact that God created the universe and everything within it is a fundamental belief of Christianity called Creation.

Can the Creator be the same thing as the creation?

No. The distinction between Creator and creation is important to understanding the Christian worldview. Christianity holds that God and the universe are distinct, and not the same. God existed before any of the created universe, and if the entire universe were to be destroyed, or to decay, or to disappear, God will remain.

What does it mean to believe that the universe and God are the same?

Pantheism (meaning "all is God") is the belief that God and the contents of the universe are one and the same. To a pantheist, a rock is God, a tree is God, you are God, the sun is God.

Does pantheism conflict with Christianity?

Yes. Pantheism conflicts with the Christian worldview. The Christian God of Creation is separate and distinct from all His creations. In the Christian worldview, God created—from nothing—everything in the universe.

What is the Latin term for God creating the universe from nothing?

The Latin phrase "ex nihilo" means "out of nothing." Christianity teaches that God created the universe "ex nihilo." The substance and material of the entire universe is not eternal; it did not always exist. Through a creative act, God willed the universe into existence, from nothing, from no original building blocks. God did not "craft" the universe out of existing matter (as described in other non-Christian, non-Abrahamic creation myths). Remarkably, Creation from nothing ("ex nihilo") is confirmed by the scientific evidence of the Big Bang.

How can we think rationally about what or who could create everything out of nothing? What sort of entity can create time, space, matter, and energy—the entirety of everything in the observed cosmos? Returning to an earlier question, who or what could exist before there was any time, space, matter, or energy?

Whatever or whoever created time, space, matter, and energy must have certain qualities. These required qualities give us clues about the Creator or First Cause. To exist in a state in which neither time nor space nor matter nor energy existed, this entity must transcend all those fundamental properties.

What qualities does the Creator or First Cause need to create time from nothing?

To create time itself, this creative entity must be timeless—outside of and not subject to time.

What qualities does the Creator or First Cause need to create space from nothing?

To create space from nothingness, this creative entity must be spaceless and changeless. As the First Cause of all space, the First Cause cannot be subject to the constantly changing material reality that we humans perceive, nor confined by the physical space we inhabit.

What qualities does the Creator or First Cause need to create matter and energy from nothing?

To create all the matter and energy that ever existed, this creative entity must be immeasurably and immensely powerful and immaterial.

In describing the qualities of what or who could have created time, space, matter, and energy—just as both the Big Bang and the Biblical Creation describe—what or who have we revealed?

God! We just arrived at the classic understanding of God. We revealed a creative, transcendent entity that is timeless, spaceless, changeless, immaterial, and immensely powerful. By thinking about this origination of the cosmos—under both the Christian Creation and the scientific Big Bang—we have described a creative, transcendent agent that matches the description of what we consider God.

Well, then who created God?

God, the Creator, the First Cause, is not and cannot be subject to the limitations of His creations, such as time, space, matter, and energy. As the Creator and First Cause, God transcends time, space, matter, and energy. Therefore, God has no "cause" or source of creation for Himself.

Why is it important to understand that observed evidence of the Big Bang demonstrates the entire universe had a definite beginning?

The finite, rather than eternal, universe provides strong reasons to believe in God the Creator. If the universe were eternal, there would be no Big Bang or Creation. With no Big Bang or Creation, there is no reason to ask, "What caused the Big Bang?" or "Who created the universe?"

Throughout history, mankind has asked whether the physical universe we inhabit is eternal or finite. Did the universe always exist, or did it come into being at a certain point in time? Scientific observation answered this question in the 20th century. The universe came into being, the universe had a beginning; our cosmos is not eternal.

What is the scientific evidence for the beginning or creation of the universe?

Four separate lines of scientific evidence independently support the beginning or creation of the universe known as the Big Bang. (1) Scientists observe distant galaxies moving quickly away from our galaxy, confirming an expansion of the entire universe from a single, initial point. (2) Scientists measure the distribution of elements throughout the universe in precisely the same proportions expected by these elements proceeding from the Big Bang. (3) Scientists observe cosmic microwave background radiation throughout the universe in accordance with the tremendous release of radiation during the Big Bang. (4) Scientists map out the universe in precisely the large-scale structure and formation of galaxies expected by a cooling of temperatures of the universe some 10,000 years after the Big Bang, establishing matter as the dominant form of energy density. All four of these scientific observations or measurements are independent of one another and all four support the theory of the Big Bang.

How does the scientific evidence of the finitude and beginning ("Big Bang") of the universe support arguments for the existence of God?

If the universe is eternal, it has always existed. Something that has always existed does not need a cause attributed because it has always existed. Knowing that the universe suddenly came into being from nothing implies a cause. The Christian explanation for this cause is God, the Creator of the universe from nothing.

Furthermore, the fact that the universe had a beginning and is not eternal proves the second premise of the Kalam Cosmological Argument, one of many compelling arguments for the existence of God.

What is the Kalam Cosmological Argument?

The Kalam Cosmological Argument for the existence of God was revived by the Muslim scientist and theologian al-Ghazali in the 11th century. His argument has two premises and a conclusion: (1) Anything that comes into being has a cause. (2) The universe came into being. Therefore, the universe has a cause. As you just read, both Christianity and the best scientific evidence affirm the second premise: (2) The universe came into being. The first premise is axiomatic (self-evident, unquestionable). Logically, the conclusion is that the universe has a cause. If you have a better explanation for the cause of the universe than God, you really need to write a book!

8 Chance, Necessity, or Design? Our Fine-Tuned Universe

Is our universe the product of chance, necessity, or design? The physical properties of our cosmos are contingent, not necessary – they could have been arranged differently. Our cosmos could not have ordered itself by chance, the probability calculations are not even close. By eliminating chance and necessity, the best explanation is design, which is also confirmed by the incredibly fine-tuned constants, equations, and physical properties of the cosmos. Even the slightest change to any of several physical laws of the cosmos erases the possibility of any life within ordered galaxies with stars and planets. Compelling evidence of design is revealed within the fundamental scientific structure of our finely tuned cosmos.

What are the three possible explanations for the origin and nature of our fine-tuned cosmos?

The origin and nature of our fine-tuned universe and everything within it can be explained in one of three ways: random chance, necessity, and design by an intelligent Creator. Random chance would explain everything as a coincidence or accident and "it just happened this way." Necessity would explain that everything had to happen as it did. In other words, our universe exhibits these special, fine-tuned properties because no other configurations are possible and "it had to happen this way."

The third explanation is design. Someone or something designed the universe and its properties to sustain life and life-giving formations like suns, planets, galaxies, and stable elements of chemistry. This chapter will show that random chance and necessity cannot explain the origin and nature of our cosmos, leaving only the design explanation of God as the Creator, origin, and designer of our universe.

Can random chance explain the origin and nature of our cosmos?

No. The unrealistic theory behind random chance is that the "luck of the draw" can explain that a universe like ours could arise with all its finely tuned physical properties, laws, fundamentals, and constants that allow for life and the ordered cosmos as we know it. When we examine the physical properties of the universe, it becomes overwhelmingly obvious that mere random chance cannot account for the extreme fine-tuning and improbably balanced relationships.

The laws of physics are unquestionably "dialed in" or specified in such a narrow range, and in such an impressive series of combinations and connections, that random chance is not realistic. When physicists and mathematicians calculate the probability of the fine-tuning of our universe by random chance, the fractional number is so tiny that you could spend your entire lifetime writing out the number of the denominator and never come close to one percent completion. Even if you extended your lifetime to 14 billion years—beyond the age of the cosmos—you would never come close.

What is the mathematical probability of our cosmos arising by mere chance?

Focusing on just one physical parameter of many finely tuned parameters, Nobel laureate and physicist Roger Penrose (born 1931) calculated that the odds of our cosmos arising by mere chance is one in ten to the tenth power—to the tenth power—to the 123rd power (1/10^10^23). To put this mind-boggling number in perspective, the best estimate is that there are ten to the 79th power atoms in all the cosmos, across all galaxies, including the entire expanding universe. There are also a multitude of other such highly improbable calculations for the other fundamental forces and constants, which we will explore in the next chapters. Therefore, it is not reasonable to say that the cosmos "just happened this way" by chance.

Can necessity explain the origin and nature of our cosmos? Did our universe and its physical parameters "have to happen this way" or could it have been different?

No. Necessity fails as an explanation for several reasons. The easiest and most obvious reason necessity fails is due to the fundamental constants or values. When the laws and parameters of physics are expressed in mathematics, certain values remain constant, like the force of gravity. These constant values do not depend on the laws of physics because other values for the force of gravity could also be compatible with the same laws of physics. In this way, the constants are arbitrary, could be different, and are therefore not by necessity.

The same arbitrariness holds true for the initial conditions of the universe as expressed by quantities of initial matter and energy, and by the initial ratio between order and disorder (called entropy). The initial conditions of the universe are also extremely fine-tuned, but these values and ratios could have been different. If the initial conditions were even slightly changed, our cosmos as we know it could not exist nor could life. Yet physicists confirm that these initial conditions could have been different. Therefore, necessity fails as an explanation for the origin and nature of our cosmos because "it did not have to happen this way."

What is natural theology?

Natural theology uses human reason and observation about the natural world to understand God.

What is an example of using natural theology to explain the origin of the cosmos?

Observing that the universe is finely tuned for the existence of life points to a personal, intelligent designer that ordered and created the universe and everything in it. By proving that the circumstances, conditions, and laws of physics necessary for life and our cosmos are infinitesimally unlikely based on chance alone, and could have been set to different values, we can infer to the best and only remaining explanation of design.

How can natural theology rely on scientific observation, knowledge of physics, and calculations of mathematical probability to infer a hypothesis of an intelligent designer?

The ordered universe can be explained by chance, by necessity, or by a designer. Since we know that the ordered universe is not explained by necessity (since it is possible that the laws and constants of physics could be different), nor by chance (as demonstrated by scientific and mathematical calculations), the inference to the best explanation is that of an intelligent designer who set constraints with the purpose of creating a universe like ours that sustains life. The intelligent designer of the cosmos is commonly called "God."

What is revealed theology?

Revealed theology relies upon what God has communicated with humanity to understand God.

What is an example of using revealed theology to explain the origin of the cosmos?

Relying on the Creation account within the book of Genesis in the Old Testament of the Bible to explain how our universe came into being is an example of revealed theology. The Bible is the paramount example of revelation, or knowledge imparted from God. Reading the Bible does not require additional observations about the natural world. Instead, the words of Scripture provide revealed knowledge of God (revealed theology). For example, Genesis, chapter 1 explains that God created the universe.

Can revealed theology and natural theology arrive at the same conclusion about the origin of the cosmos?

Yes. It is possible for the strikingly different methods of reasoning found in natural theology versus revealed theology to arrive at the same conclusion. If those divergent methods of reasoning arrive at the same conclusion, such independent confirmation is persuasive evidence of the truth of the matter. For example, both natural theology and revealed theology point to an intelligent designer of the created universe.

What is the anthropic principle? What does "anthropic" mean?

"Anthropic" derives from the Greek word for human being, "anthropos." The anthropic principle is the scientific and philosophic (not religious) concept that not just our planet, but our entire universe is fine-tuned to allow for human life.

How did scientists and philosophers discover the anthropic principle?

By examining the laws of physics, fundamental forces, constants, and initial conditions, scientists, mathematicians, and philosophers have determined that only an extremely small range of these values permits life. Therefore, the anthropic principle (literally, the "human principle") suggests that the nature, conditions, and laws of the cosmos are improbably fine-tuned for human life. The next logical step is to reason that this fine-tuning is by design, as the product of an intelligent mind we call God. While randomness or necessity points to blind chance or inevitability, the complex order in our cosmos instead points to an intelligent designer, as demonstrated by the anthropic principle.

How do atheistic materialists distort the anthropic principle?

Atheistic materialists distort the anthropic principle to argue that the fine-tuned universe should be no surprise. Under this specious line of argument, our cosmos is habitable and life-sustaining as a necessity of the fact that we observe our cosmos. In other words, since we are alive to observe the cosmos, we should not be surprised that the cosmos allows us to live, no matter how fine-tuned everything appears.

What hypothetical scenario illustrates the logical fallacy of using a skeptical interpretation of the anthropic principle to downplay the importance of the fine-tuned universe?

Imagine yourself blindfolded and standing 20 feet in front of a firing squad of 40 expert marksmen. All these expert sharp shooters are aiming at you with precisely maintained rifles with calibrated scopes. After an announced countdown from ten, you hear the deafening sounds of all 40 rifles discharging bullets. Miraculously, you feel no pain. You pull off your blindfold. You examine yourself, and to your dumbfounded delight, not a single bullet struck your body.

How does the atheistic materialist interpret your survival despite the 40 expert marksmen?

The atheistic materialist would argue that it is no surprise that no bullets struck your body. You survived the ordeal; therefore, we should not be surprised that all 40 expert marksmen missed their respective shots. In the same reasoning, the incredible fine-tuning of the universe is no mystery and nothing consequential because we exist to observe the effects. If the universe were not fine-tuned, we would not exist to observe it. If any of the 40 marksmen hit the intended target, you would not be alive to notice.

What is the more reasonable interpretation of your survival despite the 40 expert marksmen?

Obviously, the more reasonable interpretation is that something anomalous happened—a conspiracy by design, not random chance. The fact that 40 expert marksmen, all at once, failed to hit a stationary target 20 feet away screams out for an explanation that cannot be explained by necessity or chance. The best inference is that the marksmen colluded amongst themselves to miss your body to save your life—a conscious conspiracy by design. In the same way, an incredibly fine-tuned cosmos is best explained by design.

9 How is the Universe Fine-Tuned?

Our cosmos is so overwhelmingly fine-tuned that even atheistic materialist physicists and mathematicians are compelled to admit the stark evidence. The degree of fine-tuned precision for a multitude of physical parameters of our universe is astonishing and undeniable.

Is the universe fine-tuned?

Yes. Our universe is incredibly fine-tuned to support life, planets, solar systems, and the marvelous interaction of matter and energy that forms, shapes, and sustains existence.

What does it mean to say the universe is fine-tuned?

The physical laws and properties of our universe are unmistakably and improbably "dialed-in" or prearranged to support life in an extraordinary combination of ways at an astonishing degree of precision.

How did physicist Paul Davies metaphorically describe the fine-tuning of the universe?

Physicist Paul Davies (born 1946) introduced a clever metaphor to describe the many ways our cosmos is fine-tuned to allow life:

"Before you is a designer machine that lets you tinker with the basics of physics. Twiddle this knob and you make all electrons a bit lighter, twiddle that one and you make gravity a bit stronger, and so on. It happens that you need to set thirtysomething knobs to fully describe the world about us. The crucial point is that some of those metaphorical knobs must be tuned very precisely, or the universe would be sterile."

How do physics and mathematics demonstrate the fine-tuning of the cosmos?

Physics is the scientific study of the nature and properties of matter and energy in our universe. Mathematics is the rational language of physics that describes our cosmos and the behavior of light, radiation, heat, sound, magnetism, mechanics, motion, gravity, and much more. When we express the laws and properties of physics in mathematics, we can calculate the precise values, ratios, relationships, forces, and equations. By observing and manipulating these numbers and equations, we have discovered not just the true or actual values, but the range of possible values.

How likely is it that our universe came into being in the way we find it purely through chance?

The probability of randomly selecting the fine-tuned values of our cosmos is so infinitesimally low that it is effectively zero, practically impossible. The overwhelming evidence says that the possible range of values for physical laws and properties is astronomically larger than the specific values of our actual universe. Yet only the tiniest subset of possible values—like the ones exhibited in our cosmos—are hospitable to life or even to a stable universe with distinct elements, particles, molecules, planets, etc.

In other words, the fundamental rules and initial conditions of our cosmos "hit the jackpot." The odds are extraordinarily narrow and hard to fathom, far exceeding by orders of magnitude the chance of winning the lottery, several times in a row. The existence of intelligent life and of an ordered universe (rather than unorganized chaos) depend on a complex and delicate balance of initial conditions present since the Big Bang some thirteen billion years ago. The reasonable conclusion is that this improbable, highly precise, fine-tuned arrangement of physical parameters is clear and convincing evidence of an intelligent designer rather than the impossible result of random chance.

What are some physical laws, properties, equations, and values that exemplify the fine-tuning of the universe?

The following is an incomplete list of the physical parameters that are incredibly fine-tuned: entropy, gravitational constant, strong nuclear force, weak nuclear force, electromagnetic force, cosmological constant or expansion speed of the universe, initial distribution of mass versus energy, ratio of protons and electrons, speed of light, and more. Note that these parameters represent the fine-tuning of the entire cosmos. In addition, our earth and solar system also exhibit the "Goldilocks principle" of having a plethora of physical conditions and ratios "just right" to support life on our planet.

10 The Fine-Tuning of Entropy, Expansion, Gravity, and the Fundamental Forces of Physics

Surveying the fundamental properties of our physical universe reveals an extensive list of astonishingly fine-tuned parameters. Whether we examine entropy, gravity, electromagnetism, nuclear attraction, the expansion rate of the cosmos – or critical ratios between the strength of these various forces – the evidence for fine-tuned precision is overwhelming. The mathematical probability of any of these fine-tuned conditions occurring by random chance is too tiny to comprehend, let alone to take seriously as a possibility.

What is entropy?

Entropy corresponds to disorder, so when entropy decreases, order increases. Your messy bedroom is in a state of high entropy. After you clean your room, you have reduced the entropy by restoring it to a higher organizational state.

Is our cosmos in a state of high entropy or low entropy?

Our cosmos, with defined planets, stars, and galaxies, is a low-entropy system. The universe exhibits a remarkably high degree of order; therefore, entropy is lower.

What would the universe look like if it were in a state of high entropy?

Chaotic. With high entropy the universe would be unrecognizable and disordered, with matter haphazardly distributed, resulting in many black holes. The random accumulation of vast amounts of matter would form black holes that would suck in and collapse any possible planets, stars, or galaxies.

How did the physicist Roger Penrose calculate the probability of our cosmos exhibiting its low entropy state that allows for the planets, stars, and galaxies to form?

Nobel laureate and physicist Roger Penrose (born 1931) calculated the sum or series of values to the maximum upper range of the highest possible state of entropy—that state which would result in a gigantic, all-encompassing black hole rather than an ordered universe with recognizable distinctions like planets or stars or galaxies. Then, Penrose calculated the probability of our low-entropy, actual cosmos, given the immense number of possible entropic combinations.

What did Roger Penrose calculate to be the probabilistic odds of our cosmos achieving its low entropy state that allows for an ordered universe of planets, stars, and galaxies?

Roger Penrose calculated the probability of the initial low-entropy state of the universe, by chance alone, is one part out of ten, to the power of ten, to the power of 123 (1/10^10^123). This number is incomprehensibly small and describing it as "extremely unlikely" does it no justice. For context, scientists have estimated that there are ten to the power of 80 (10^80) particles in the universe. Even that number of all the particles in existence is a tiny fraction of the number that Penrose calculated. Therefore, Penrose points to an inescapable conclusion: The universe did not arise in its ordered state of low entropy by random chance.

What is the fine-tuning of the expansion of the universe? What is the cosmological constant and how fine-tuned is the value observed in our cosmos?

Our universe is constantly expanding. To quantify the expansion of the cosmos, Nobel laureate and physicist Albert Einstein (1879 - 1955) discovered the cosmological constant. Einstein identified the cosmological constant as the density of space that allows for expansion balanced against the gravitational force of attraction pulling against expansion.

The cosmological constant is fine-tuned to one part out of ten to the power of 120 (1/10^120). The denominator of this tiny fraction is the number one followed by 120 zeroes—much greater than the number of particles in the entire cosmos. If the cosmological constant were slightly greater, the entire universe would have collapsed upon itself into a giant, all-encompassing black hole. Conversely, if the constant were barely decreased, the universe would not form any planets, stars, or galaxies because matter would expand too quickly into formless disorder.

What is described by the various "constants" within physical laws of the universe?

The "constants" refer to proportionate quantities of variables, like the relationship between pressure and volume, or between force and acceleration. Constants between variables can be directly proportionate or inversely proportionate. If variable x increases as y increases, there is a directly proportionate constant between x and y. On the other hand, if variable a decreases as b increases, then there is an inversely proportionate constant between a and b.

What is one of the most famous physical constants?

You may recall the gravitational constant, G, from your high-school physics class. Isaac Newton discovered that the gravitational force, G, is directly proportionate to the multiple of the mass of the two bodies in question, divided by the square of the distance between them. This proportionate relationship can be considered the level of gravitational force between objects.

The gravitational constant is finely tuned to one part in ten to the power of 34 (1/10^34). The denominator of this tiny fraction is the number one followed by 34 zeroes. If the gravitational constant were increased slightly, stars would burn out too fast, and if it decreased slightly, stars would not burn long enough to produce the elemental building blocks of life like carbon, hydrogen, oxygen, and nitrogen.

What is the expansion rate of the cosmos?

The expansion rate of the cosmos measures how fast galaxies move away from each other as the universe expands. If you draw points on a balloon with a marker, when you inflate the balloon, those points will move away from each other. Think of the expansion rate as the speed at which those points on the surface of the balloon separate, representing different galaxies of stars and planets in the entire cosmos.

How is the expansion rate of the cosmos fine-tuned?

If the cosmos expands too quickly then stars, planets, and galaxies cannot form. The force of gravity would not "hold together" or coalesce the heavenly bodies. Instead, the universe would expand into disordered, diluted matter that is too separated to connect into anything coherent. If the cosmos expands too slowly, the force of gravity would crush all the expanding matter into a giant black hole sometimes called the "big crunch."

The expansion rate of the universe is fine-tuned to one part in ten to the power of 55. The denominator of this tiny fraction is the number one followed by 55 zeroes.

What is the electromagnetic force?

The electromagnetic force causes the attraction of particles with opposite charges, like a positively charged proton and a negatively charged electron. In a similar manner, the electromagnetic force causes particles of opposite charge to be repelled away from one another, like two negatively charged electrons.

How is the electromagnetic force fine-tuned to support life in our cosmos? How fine-tuned is the ratio between the electromagnetic force and the force of gravity?

Among many other problems, if the electromagnetic force were slightly stronger, stars would burn too cold to support life. In the opposite direction, a weaker electromagnetic force would result in stars that burn too hot and too quickly to support life.

The electromagnetic force is fine-tuned to about 1 part in 25, or 4 percent. But the ratio of the electromagnetic force to the force of gravity is fine-tuned to 1 part in 10 to the power of 37 (1/10^37). The denominator of this tiny fraction is the number one followed by 37 zeroes.

What is the strong nuclear force?

The strong nuclear force causes the attraction and binding together of protons and neutrons. Protons are positively charged but neutrons have no charge (or a neutral charge). While electromagnetic force causes oppositely charged particles to attract, that force has no effect on a positively charged proton and a neutron. Therefore, the strong nuclear force is vital to the structural cohesion of atoms, the fundamental building blocks of all matter. We need the strong nuclear force to form and sustain elements necessary for life like carbon, hydrogen, oxygen, and nitrogen.

How is the strong nuclear force fine-tuned to support life in our cosmos?

To produce enough carbon to support all the carbon-based life forms on earth, the strong nuclear force must be calibrated within about 1 to 4 percent of its current level. No carbon means no life. If the strong force were reduced slightly, elements heavier than hydrogen could not form with their larger number of protons. If the strong force were increased, then hydrogen would turn into helium and other heavier elements, starving the universe of water and fuel for stars.

Even more shockingly, the ratio of the weak nuclear force to the strong nuclear force is fine-tuned to 1 part in 10,000. If this ratio of the weak and strong nuclear forces were any less "dialed in," stars could not be powered by hydrogen fusion, and could not support life on our planet or any others in the cosmos.

What is the weak nuclear force?

The weak nuclear force decays atoms, the elemental particles of matter. Through nuclear radiation, emanating from the nucleus of every atom, all atoms slowly decay at differing rates. This atomic decay creates other elements, contributing to the diversity of elements needed for life.

How is the weak nuclear force fine-tuned to support life in our cosmos?

Any slight change in the weak nuclear force would cause serious problems. If the weak nuclear force increased slightly, more rapid hydrogen fusion would create a cosmos dominated by helium, precluding the formation of water and stable star systems. A small decrease to the weak nuclear force prevents enough helium needed to generate heavy elements within stars.

How can we summarize several of the incredibly fine-tuned physical parameters of our cosmos?

Physical forces, constants, and laws of the universe govern the order, rationality, predictability, and life-sustaining properties of our cosmos. These are the "rules" for the mechanistic understanding of the universe recognized by atheistic materialists. Yet the same "rules" of materialism, physicalism, and mechanistic phenomena are incredibly fine-tuned beyond any reasonable consideration of mere chance. Instead, the "rules" appear to be carefully crafted, not random at all. The same scientific process that discovered these "rules" also confirms that they are not necessary and did not "have to happen this way." With random chance and necessity eliminated as possible explanations for the astonishingly fine-tuned cosmos, the remaining explanation is design.

When we ponder the designer of the cosmos, the discussion inevitably leads to a timeless, spaceless, changeless, immaterial, and immensely powerful entity. To be the source, creator, and designer, this entity must transcend all time, space, matter, and energy, which appeared suddenly and spontaneously from a mysterious state of nothingness at the Big Bang. In describing the improbable nature of the "rules" of our universe, the evidence for their design, and the nature of the only possible designer, we describe exactly the classic understanding of God. We should not be surprised that scientific evidence of the 20th century confirmed the centuries-old theological wisdom revealed by the first book of the Bible that "in the beginning, God created the heavens and the earth."

11 How Physicists Describe the Fine-Tuning of the Universe

Even atheistic materialist scientists have no choice but to admit that our cosmos is incredibly fine-tuned, eliminating the possibility of our universe arriving by random chance. The term "Big Bang" was coined by an atheist (Fred Hoyle) who was disappointed that cosmologists confirmed the religious belief that the universe had a distinct, sudden beginning rather than eternal existence. The 20th century was awkward for atheist astronomers because the ancient debate of the finitude of the cosmos was settled in favor of Christian theology. Even worse for the atheist worldview, many of the fundamental physical properties of the universe were discovered to be fine-tuned within an incredible degree of precision.

What did physicist Stephen Hawking say about the fine-tuning of the fundamental constants and ratios of physics?

Physicist and confirmed atheist Stephen Hawking (1942 - 2018) stated in his 1988 book, A Brief History of Time, that the evidence for our "very finely adjusted" cosmos is compelling:

"The remarkable fact is that the values of these numbers seem to have been very finely adjusted to make possible the development of life."

What did the astronomer Fred Hoyle say about the discovery of scores of examples of the exceedingly improbable fine-tuning of the universe?

Atheist astronomer Fred Hoyle (1915 - 2001) stated in 1981:

"A common-sense interpretation of the facts suggests that a super-intellect has monkeyed with physics, as well as with chemistry and biology, and that there are no blind forces worth speaking about in nature. The numbers one calculates from the facts seem to me so overwhelming as to put this conclusion almost beyond question."

Notably, Hoyle is the same atheist astronomer who in 1949 coined the term "Big Bang" as an insult due his disappointment at the confirmation that the universe had a distinct beginning. As a committed atheist, Hoyle had hoped that the cosmos was eternal and therefore had no need to explain its origin. Years later in the quoted interview, Hoyle had to come to terms with both the finite beginning of the universe as well as its unmistakably fine-tuned properties—difficult pills to swallow for the atheistic materialist worldview.

How did biochemist Michael Denton explain the evidence of design and purpose found within the fine-tuning of the cosmos?

Biochemist Michael Denton (born 1943) described in his 2002 book, Nature's Destiny: How the Laws of Biology Reveal Purpose in the Universe, that the fine-tuning of the laws of nature points to design and purpose (teleology is the study of purpose favored by Aristotle):

"Four centuries after the scientific revolution apparently destroyed irretrievably man's special place in the universe, banished Aristotle, and rendered teleological speculation obsolete, the relentless stream of discovery has turned dramatically in favor of teleology and design..."

What did Nobel Prize-winning physicist Charles Townes say about the fine-tuning of the cosmos?

Nobel laureate physicist Charles H. Townes (1915 - 2015) confirmed and marveled at the fine-tuning of the cosmos as follows:

"This is a very special universe: it's remarkable that it came out just this way. If the laws of physics weren't just the way they are, we couldn't be here at all. The sun couldn't be there, the laws of gravity and nuclear laws and magnetic theory, quantum mechanics, and so on have to be just the way they are for us to be here."

What did physicist Paul Davies say about the fine-tuning of the fundamental forces of physics?

Physicist Paul Davies (born 1946) has commented in many ways about the implications of the fine-tuning of the cosmos:

"There is for me powerful evidence that there is something going on behind it all...It seems as though somebody has fine-tuned nature's numbers to make the Universe...The impression of design is overwhelming."

"The really amazing thing is not that life on earth is balanced on a knife-edge, but that the entire universe is balanced on a knife-edge, and would be total chaos if any of the natural 'constants' were off even slightly."

"If physics is the product of design, the universe must have a purpose, and the evidence of modern physics suggests strongly that the purpose includes us."

What did physicist and mathematician Freeman Dyson say about the human-friendly fine-tuning of the universe?

Renowned physicist and mathematician Freeman Dyson (1923 - 2020) confided that his scientific studies led him to believe that:

"The more I examine the universe and study the details of its architecture, the more evidence I find that the universe in some sense must have known that we were coming."

How did astronomer and cosmologist Edward R. Harrison interpret the evidence of a fine-tuned universe?

Astronomer and cosmologist Edward R. Harrison (1919 - 2007) declared that the fine-tuned universe is cosmological proof of design by God:

"Here is the cosmological proof of the existence of God—the design argument of [18th-century British philosopher William] Paley—updated and refurbished. The fine tuning of the universe provides prima facie evidence of deistic design."

How did former atheist Frank Tipler describe the impact of his scientific studies on his theological and cosmological worldview?

Physicist and cosmologist Frank Tipler (born 1947) was an atheist until the unmistakable facts revealed by his scientific studies forced him to believe in God as the designer of the universe:

"When I began my career as a cosmologist some twenty years ago, I was a convinced atheist. I never in my wildest dreams imagined that one day I would be writing a book purporting to show that the central claims of Judeo-Christian theology are in fact true, that these claims are straightforward deductions of the laws of physics as we now understand them. I have been forced into these conclusions by the inexorable logic of my own special branch of physics."

Why do many scientists and academics keep quiet about the deeper scientific, theological, and philosophical implications of the Big Bang and fine-tuned universe?

Atheistic materialism is the unspoken rule for many in the scientific community. Peer review, "publish or perish," academic tenure, and immense social pressure against any non-material or religious explanations combine to suppress, repress, deny, isolate, discourage, and prevent the honest, open, earnest discussion of scientific issues—when they conflict with the dominant worldview of atheistic materialism.

12 The Multiverse and Ockham's Razor

Our fine-tuned cosmos is so devastating to the worldview of atheistic materialism that some propose the Multiverse, a fantastic ensemble of alternate universes (never detected, just imagined). To dilute the importance and profundity of our fine-tuned cosmos, atheistic materialists summon infinity other realities, none of which have a shred of evidence to support their existence. The so-called Multiverse is a laughable and evidence-free confession that the atheistic materialist worldview has no credible answer for the fine-tuned universe. To avoid the conclusion of design, atheistic materialists dream up...entirely imagined universes...in a feeble attempt to rewrite the probability calculations. So much for the Multiverse.

How does the atheistic materialist respond to the startling and overwhelming evidence of the fine-tuned universe?

The Multiverse: Since our universe is fine-tuned and therefore improbably special, atheistic materialists dream up the evidence-free, fantastical, and eye-rolling idea that there are billions, trillions, or infinity other universes. Just like that, with one silly hypothetical ripped straight out of science fiction, our cosmos does not seem so special anymore.

Fine-tuning provides unwelcome conclusions for the materialist or atheist. The fact that our cosmos is fine-tuned to support life indicates a special purpose or meaning for humanity. Going further, an ordered universe with finely tuned physical conditions suggests design and the most obvious candidate for this mystery designer is God. If the material universe was created by God, an immaterial entity transcending the physical realm, both atheism and materialism are firmly refuted. So, to dodge the implications of the fine-tuned universe, atheistic materialists propose the Multiverse theory.

What is the Multiverse theory? What is the etymology of the Multiverse?

The Multiverse theory imagines an infinite number of realities, perhaps existing in other dimensions. Our reality and universe is not unique or special, but rather one droplet of water in an endless ocean of other realities and universes (described collectively as the Multiverse). "Multiverse" is a blend (portmanteau) of the words "multiple" and "universe." If you believe in the Multiverse, you believe in multiple universes yet to be discovered, not just ours.

Is there any evidence for the Multiverse theory?

No. Nothing. Not a shred of evidence exists to support the Multiverse. Rather, the Multiverse theory is a convenient fantasy that ducks the question of why our universe is fine-tuned to support our existence, the order and rationality of the cosmos, and all known lifeforms.

Does the Multiverse theory simplify anything?

No. The Multiverse theory complicates the mystery. Instead of the need to explain a single universe—which is difficult enough—the Multiverse theory requires explanation of multiple or infinite universes beyond our own. We cannot access or perceive these alternate universes, nor do we have any way to confirm their existence.

Is the Multiverse theory the same concept employed by comic books and movies about Spider-Man, the X-Men, and the Avengers from the Marvel Cinematic Universe and Marvel Comics?

Yes. Indeed, the idea of the Multiverse was heartily adopted by comic book writers and creators like Stan Lee (1922 - 2018) in the 1960s. Talented and imaginative comic book storytellers recognized the fictional, flexible, and fantastical power behind the ability to dream up an infinite number of alternate universes.

Why settle for selling one version of Spider-Man when you can sell one thousand versions? Why worry about maintaining editorial and narrative consistency in one universe when you can spread storylines across the Multiverse? These questions about the comic book version of the Multiverse have poetic relevance to the so-called "scientific" proposals about the Multiverse.

What is Ockham's razor? Who is William of Ockham?

William of Ockham (1287 - 1347) was a philosopher and theologian in the 14th century. Also known as the principle of parsimony, Ockham's razor is a problem-solving tool that favors explanations with fewer parts or assumptions. Following Ockham's razor, the simplest explanation is often the best. William of Ockham specifically warned that "Entities must not be multiplied beyond necessity" and "Never posit pluralities without necessity." The prudent advice is to minimize, rather than multiply, explanations and explanatory steps.

Does the Multiverse theory violate Ockham's razor?

Yes, egregiously so. William of Ockham's razor warns against excess multiplication of explanatory entities. The Multiverse theory introduces multiple entities, but these entities are especially complicated entities, because each one is an entirely different universe or dimension. Usually, a violation of Ockham's razor involves one extra assumption, person, mechanism, or principle. Yet the Multiverse theory imagines entire universes that each cry out for their own detailed explanations. In this way, the Multiverse theory directly contradicts Ockham's razor.

What is the gambler's fallacy?

The gambler's fallacy is a mistake in reasoning whereby you think that a random event is somehow "due" to occur—more likely than as predicted by probability—simply because of how past events occurred. For example, many gamblers mistakenly believe that if you flip a coin and it lands on the same side (heads) a few times in a row, you should bet money on the coin landing on the opposite side (tails). The incorrect reasoning is that the series of flips landing on heads somehow influences the next flip of the coin, and therefore the next coin flip is unusually more likely to land on tails. Of course, this is incorrect because every time you flip the coin, there is a 50 percent chance of landing on heads or tails, irrespective of past events.

What is the inverse gambler's fallacy?

The inverse gambler's fallacy is a mistake in reasoning whereby you think that an unlikely outcome of a process means that process must have happened many times in the past. For example, you see someone roll a pair of dice into "Snake Eyes" twice in a row. You commit the inverse gambler's fallacy by incorrectly assuming the dice must have been rolled many times before to finally arrive at the lucky streak.

In coin flips and dice rolls, the odds remain the same each time. In this way, the traditional and inverse gambler's fallacy commits the same fallacy: thinking that one probabilistic event is connected to other, independent, unrelated probabilistic events.

Does the Multiverse hypothesis commit the inverse gambler's fallacy?

Yes. The Multiverse hypothesis is a prime example of committing the inverse gambler's fallacy. The Multiverse hypothesis imagines untold trillions of other realities simply because our fine-tuned reality is so unlikely or "lucky." Atheistic materialists recognize the incredible unlikelihood of our fine-tuned universe coming into existence yet want to deny the obvious implications thereof. So, committing the inverse gambler's fallacy by imagining infinite multiverses is a clumsy way to ignore the troublesome line of inquiry before others draw obvious conclusions about the fine-tuning and design of the cosmos.

13 Materialism vs. Idealism

Materialists believe the fundamental nature of reality is tangible matter, while idealists hold that reality is based on immaterial minds. An expansive "middle ground" of worldviews exists between the polar extremes of materialism and idealism. Yet adopting strict materialism is untenable because the immaterial mind, not unconscious matter, is at the heart of all scientific understanding. Furthermore, if human minds are nothing more than material brains deterministically assembled by random variation, there is no reason to take seriously anything ever thought or said by anyone – an idea that Charles Darwin called his "horrid doubt." The debate between idealism and materialism touches upon mind and matter, cosmos and consciousness.

In the fundamental nature of all reality, is everything more like physical matter or more like immaterial minds? What arrived first: mind or matter? What ultimately controls change, cause, and effect? Is it mind or matter that accomplishes anything?

You just stumbled into the timeless debate between idealism and materialism, two extremely opposite worldviews.

What is idealism?

Idealism means that in the ultimate reality of existence, immaterial mind is the supreme and primary source. Mind came before matter. Mind is superior to matter. Idealism is a worldview of "mind over matter" that emphasizes the immaterial aspects of reality.

What is materialism? What is atheistic materialism?

Materialism views all reality as mechanical: atoms and molecules crashing about due to physical forces, readily verifiable by the scientific method. Going further, the atheistic materialist rejects the existence of spiritual entities like the human soul, angels, demons, and God. Spiritual concepts like divinity, redemption, salvation, afterlife, worship, and transcendence are all summarily rejected as fantasy, lies, superstition, and nonsense, totally unrelated to verifiable truth.

What is the fundamental nature of reality according to the atheistic materialist?

The atheistic materialist believes that our reality, our entire cosmos, our very being, can be reduced to physical matter. The very name "materialist" invokes this metaphysical claim. Matter is the beginning and the end, and nothing exists beyond matter.

What is the deepest or most precise understanding of our existence according to the materialist?

To the materialist, atoms, subatomic particles, or quantum particles represent the most precise understanding of our existence. These quantum particles are simply the smallest units of matter, and the reductive materialist believes that matter represents the entirety of existence. The indivisible building blocks of matter represent the metaphysical limits of knowledge.

Who or what profession is best positioned to explain and describe our existence, according to the materialist? Is that profession known for having an intellectual and philosophical bias toward materialism?

If everything is reduced to matter, and all truth relates to physical observations of the natural world, then the scientist is best positioned to explain everything. Indeed, many modern scientists adopt the materialist worldview. This circular, self-affirming connection deserves attention. Adopting a materialist worldview restricts knowledge and truth to that which is confirmed by scientific observation, and many scientists prefer the very same worldview that elevates their chosen profession. One hand washes the other. You may also notice that traditional allopathic doctors are usually skeptical of alternative and holistic medicine.

Can materialism support its own basic claims?

No. Materialism cannot support its own basic claims. By denying the existence of truths outside scientific confirmation, materialism refutes itself. The claim that no truth or reality exists outside of matter is impossible to prove by scientific methods. The materialist outlook also denies the reality of conscious minds, relegating them to ephemeral or illusory phantasms of the physical brain. Therefore, materialism cannot support any intellectual propositions, at least not any proposed by the mind of a human being.

What did the originator of the materialist theory of evolution, Charles Darwin, say about the "horrid doubt" and difficulty in trusting human thoughts if humans are simply evolved animals and consist exclusively of matter?

Charles Darwin (1809 - 1882) himself, originator of the materialist theory of evolution, immediately recognized the epistemological limitation of his theory when combined with atheistic materialism. If humans are nothing but matter, then human brains are also matter. If humans are nothing more than evolved apes—intelligent animals—then the thoughts produced by human minds deserve no more confidence than the hooting of howler monkeys. In 1881, Darwin stated his "horrid doubt" in a letter to William Graham:

"But then with me the horrid doubt always arises whether the convictions of man's mind, which has been developed from the mind of the lower animals, are of any value or at all trustworthy. Would any one trust in the convictions of a monkey's mind, if there are any convictions in such a mind?"

How did the physicist John Polkinghorne describe the challenge of believing any thoughts at all if brains are nothing more than matter? Why should we trust the product of unconscious brain tissue encased in the human skull, supposedly derived only from natural selection of random variation from lower-order animals?

British physicist John Polkinghorne (1930 - 2021) stated in his 1986 book, One World: The Interaction of Science and Theology:

"Thought is replaced by electro-chemical neural events. Two such events cannot confront each other in rational discourse. They are neither right nor wrong. They simply happen. The very assertions of the reductionist himself are nothing but blips in the neural network of his brain. The world of rational discourse dissolves into the absurd chatter of firing synapses. Quite frankly, that cannot be right and none of us believes it to be so."

Is there an established term for the fundamental challenge and contradiction of the existence of our minds and consciousness when considered within the materialist worldview?

Yes. The Hard Problem of Consciousness is the established term to describe the inability to explain consciousness and minds emerging from unconscious matter within brain tissue.

Why is it called The Hard Problem of Consciousness? How can consciousness emerge from unconscious matter in a materialist worldview?

The Hard Problem of Consciousness earned its name among neuroscientists, philosophers, and evolutionary biologists because the problem has never been solved and continues to stump strict materialists despite decades or centuries of contemplation. Under strict materialism, there is no coherent answer for how conscious minds emerge from unconscious matter. Australian philosopher David Chalmers (born 1966) introduced the term in his 1995 article, "Facing Up to the Problem of Consciousness."

What is one invalid yet commonly used excuse for how immaterial minds and consciousness emerge from matter? Why is it incorrect or mistaken?

Some materialists try to solve The Hard Problem of Consciousness by denying the reality of consciousness. Instead of grappling with The Hard Problem, strict materialists believe that consciousness and the perception of minds are simply illusions, and therefore, can be ignored rather than explained. The error in this line of thinking is that illusion is a form of consciousness. For perception to be distorted into false perception (illusion), there must be a perceiver experiencing the perception, thereby proving consciousness.

What is an innovative, radical, yet ultimately unsatisfying and unwieldy way to explain The Hard Problem of Consciousness?

One radical explanation for The Hard Problem of Consciousness is called panpsychism, which holds that consciousness is not unique to human experience. Instead, consciousness is infused within every particle of matter throughout the cosmos. Therefore, no special explanation for human consciousness is required because all physical matter contains consciousness.

What is unsatisfying and unwieldy about the panpsychist explanation of The Hard Problem of Consciousness?

The panpsychist explanation dilutes or multiplies The Hard Problem of Consciousness. Human consciousness is no longer special; consciousness is diluted and expanded to every atom in the universe. Consciousness itself is multiplied into enormous quantity and scope, encompassing every part of the cosmos, not just human minds but also non-living, inorganic matter like stones or water molecules.

By exploding consciousness into every particle in the cosmos, panpsychism tries to dilute its importance. Yet the panpsychist theory is extravagantly speculative, multiplying entities and explanations in flagrant violation of Ockham's razor.

How does panpsychism "explaining" consciousness relate to the Multiverse "explaining" the fine-tuned universe?

Panpsychism and the Multiverse are two exceptionally eyebrow-raising "explanations" that expose serious limitations in adopting the atheistic materialist worldview, which fails to explain the emergence of minds and of our entire cosmos. Both panpsychism and the Multiverse are extravagant speculations that violate Ockham's razor by introducing extra entities and explanations in mindbogglingly enormous quantities. Both theories also try to explain away mystery by infinitely multiplying the mystery.

If the emergence of consciousness is inexplicable under atheistic materialism, panpsychism dilutes and expands the mystery into every particle. If the fine-tuning of the physical laws, forces, and constants of our cosmos screams of careful design or at least eliminates the possibility of being random, the Multiverse theory contends that our universe is neither improbable nor special by imagining infinity other universes.

14 Cartesian Dualism and the Grand Bargain of Immaterial Mind and Matter

Rene Descartes divided the cosmos into unconscious matter and conscious, immaterial minds (or souls). The influence of Cartesian dualism of immaterial mind and matter persists today. The first step of the Cartesian "Grand Bargain" removed the study of God, spirits, and souls from scientific exploration, relegating those immaterial concepts to theology and philosophy. Later these immaterial concepts were banished entirely from academic study by the limited, myopic worldview of atheistic materialism. Despite these efforts to ignore immaterial concepts, the same problems identified by Descartes remain just as mysterious today for the atheistic materialist: How can immaterial minds and consciousness emerge from unconscious matter? What is the nature of the mind-brain relationship?

Who is Rene Descartes?

Frenchman Rene Descartes (1596 - 1650) was a 17th century philosopher, scientist, and mathematician whose first principle, "I think, therefore I am" survives as a popular phrase to this day.

What did Rene Descartes mean by the Latin phrase, "cogito, ergo sum" ("I think, therefore I am")?

To Descartes, the recognition of thought itself proves the reality of the mind and self. For thought to occur, there must be a thinker. Although skeptics may doubt the entirety of existence, claiming that all reality is an illusion, Descartes counters that doubt itself reveals a mind ("I am") that forms the doubt ("I think"). Questioning existence, pondering the mysteries of reality, proves that the mind and self, at least, exist. Without the mind and self, the questions cannot manifest. Descartes is the father of modern philosophy due to his innovative epistemology.

What is epistemology?

Epistemology is the study of "how we know what we know" or the "theory of knowledge." Epistemology explores the origin, nature, and limitations of knowledge, diving into topics like memory, perception, reason, logic, and metaphysics. Two opposing branches of epistemology are rationalism and empiricism. Descartes was a rationalist.

What is rationalism? How does rationalism relate to mathematics?

The rationalist retreats to the realm of the mind to pursue truth, believing that internal thought is superior to outside stimuli or data from the physical world. Rationalism is a theory of knowledge (epistemology) that elevates reason above other sources of knowledge like faith, sensation, or tradition.

To a philosophical rationalist, human reason is the best way to grasp fundamental truth about reality. Early rationalists were impressed by the success and precision of mathematics in describing the physical world through laws and equations of physics. Mathematics reveals numerical equations that correspond to physical reality, yet mathematics consists of abstract axioms and concepts. Naturally, the revelatory rationalism of mathematics lent authority to the idea that pure reason provides better insight than other ways of knowing. To the rationalist, important truths can be discovered solely within the human mind, untainted by external investigation, existing within the intangible realm of thought.

How is rationalism related to Pythagoras?

Both the ancient Greek polymath Pythagoras (571 - 490 BC) and the rationalist believe that the truth of all reality and knowledge lies not in the physical, tangible world, but instead within the incorporeal realm of the mind. Both theories of knowledge (epistemology) recognize mathematics and numbers as revealers of reality. Pythagoras taught that numbers are the keys to unlocking all knowledge. By focusing on an abstract, internal, immaterial concept like numbers as the basis of reality, Pythagoras was an early rationalist who interpreted the link between immaterial minds, immaterial mathematics, and the material cosmos.

What is empiricism?

Empiricism is a theory of knowledge (epistemology) that prioritizes sensory experience and experimental evidence. Empiricists oppose the method of arriving at truth by the mind alone (through rationalism) because of the inherent limitations, biases, and misperceptions of humanity. Instead, empiricists expect truth to be confirmed by recordable, testable data based on observations and measurements of the external world outside the mind. Empiricism is inherently linked to the scientific method, whereby hypotheses (ideas or possible truths) are tested by experimentation within the material world. To the empiricist, truth is found outside the mind, within tangible reality.

What is Cartesian dualism in relation to human beings?

Rene Descartes awkwardly split the human being into two realms of existence that can (in some mysterious, unexplained way) interact but are nonetheless distinct: the immaterial substance of mind and the tangible substance of matter (like the brain and body). The mystery of the mind-brain interaction, or the soul-body interaction, is heavily influenced by Cartesian dualism.

What is Cartesian dualism in relation to the metaphysics of all reality?

Cartesian dualism also refers to the metaphysical split of all reality into the immaterial (minds, souls, spirits, angels, God) and the material (unconscious matter that forms objects, animals, and human bodies). Descartes divorced the immaterial world of God, angels, souls, minds, and spirits from the material substances: the tangible, concrete, physical world of molecules, cells, organisms, human bodies, animals, and even inorganic objects like stones or tables.

How did Descartes expand the mechanical metaphor of the cosmos to include animals and living organisms?

Descartes believed that anything lacking an immaterial soul (or mind) was more like a machine, belonging to the category of mere material substance. According to Descartes, animals had no soul (or mind) and were therefore mindless automatons or machines, reacting purely from physical stimulus. While earlier philosophers and scientists imagined the planets and stars and cosmos operating like clockwork machines, following regular, mechanistic laws, Descartes expanded this worldview to include living organisms (other than human beings) as mere machines, consisting of unconscious matter. Cartesian philosophy mechanized the natural world and all living creatures except for those with minds and souls like mankind.

What is the etymology of the word animal?

The word animal comes from the Latin word "anima," meaning soul or spirit, life or breath. Therefore, the word we use for animals is an artifact of the earlier way we thought about animals as living, breathing, ensouled creatures. The atheistic materialist, who views the universe and all living creatures as mechanical machines, may be surprised to learn that the history of the word for animal evokes the concept of the soul, spirit, or life force.

How does Cartesian dualism pose a metaphysical challenge to understanding human experience?

Cartesian dualism asserts a metaphysical distinction between mind and matter. As the experience of every person involves the exercise of both the mind and body, this Cartesian separatism presents the metaphysical challenge of understanding how the mind and body interact at all, given their classification in different realms of existence.

How does Cartesian dualism reduce living organisms to machines?

For Descartes, the human body is completely material, composed of unconscious matter. Descartes considered the human body, all animals, all organisms, and all inanimate objects to be in the same category of unconscious material, relegating the entire living world to something like machines. Animals are mindless, thoughtless, emotionless machines reacting to outside stimulus. Humans differ from animals only as far as the human mind (or soul) controls the machine of the human body.

To Descartes, the mind (or soul) could be considered a type of "pilot" or controller that guides the otherwise lifeless husk of a body. Because of the Cartesian dualism wherein the mind (or soul) is completely immaterial, separate from the realm of matter, others proposed the metaphor of the "ghost in the machine" to describe the mind-body interaction problem introduced by Descartes.

What is the "ghost in the machine" metaphor and who first proposed the idea?

The philosopher Gilbert Ryle (1900 - 1976) in his 1949 book, <u>Concept of the Mind</u>, first proposed the metaphor of the "ghost in the machine" to describe (and criticize) Cartesian dualism. The ghost is the immaterial mind that "pilots" or controls the material body.

Under Cartesian dualism, can the immaterial mind and material body interact?

According to Descartes himself, the mind can signal the body to move, like when you decide to wave your hand to greet a friend. Conversely, when you place your hand in a roaring flame, your material body signals pain or sensory information to your mind.

Yet Cartesian dualism presents a serious metaphysical problem: the immaterial mind should have no engagement or ability to interact with physical matter, like the human body. By divorcing mind and matter into different metaphysical realms, the Cartesian mystery of the mind-body interaction, or the connection between the immaterial and material, continues to pose philosophical problems for modern philosophers and neuroscientists.

For Descartes, what bridged the chasm between immaterial mind and material body?

To solve his awkward problem of substance dualism, Descartes thought that there must be a central, physical location that combined or merged the sensory perceptions—like pain from your hand—before the immaterial mind could apprehend or understand those material signals.

Why did Descartes expect a central location to process or combine the material signals before the immaterial mind could access them?

For Descartes, the immaterial mind and material body occupy two separate realms of existence or reality. For those separate realms to interact, Descartes assumed the need for a device or organ that collects, combines, and translates the material sensations for the immaterial mind to perceive them. You may consider the analogy of a language translator (interpreter) serving between two diplomats who speak different languages. Perhaps a better analogy is that of the transducer, which converts physical properties into electrical signals for electronic products. To bridge the gap created by Cartesian dualism, something must stand between the immaterial mind and material body to allow two-way communication.

What did Descartes suppose was the central translator (or interpreter) between the immaterial mind and material body?

The pineal gland in the brain was thought to link the immaterial mind and material body. Descartes noticed the human brain is mostly symmetrical, sharing similar structures on the right and left sides of the brain. Yet the pineal gland is found in the center of the brain, and therefore not duplicated on both sides. For Descartes, this unique central position cast the pineal gland as the best candidate for the location of the central translator between the immaterial mind and material body.

Was Descartes correct about the pineal gland? What do modern neuroscientists think is the purpose or function of the pineal gland?

Descartes was almost certainly incorrect to think that the pineal gland interprets signals between the material body and immaterial mind. Instead, modern neuroscientists think that the pineal gland regulates the secretion of melatonin, a hormone that influences the sleep and wake cycle.

What is the Cartesian bargain? How did Descartes divide the cosmos according to his theory of metaphysics?

The Cartesian bargain refers to the cleavage established by Descartes between the material world of living bodies, nonliving objects, and other matter versus the immaterial world of minds, souls, angels, and God. After Descartes, the immaterial or spiritual world was largely excluded from science and confined to theology and philosophy. Thus began the "Cartesian bargain."

In this exchange or bargain, science gets authority over the entirety of the material universe: all living bodies, particles, plants, animals, planets, stars, and galaxies—everything in nature. Religion, philosophy, and theology is left with the immaterial remainder of the bounty, consisting of minds, souls, angels, and God.

Although the immaterial world—banished from scientific pursuits by the Cartesian bargain—is manifestly important, one cannot escape the profound loss of intellectual and philosophical territory in abdicating the entire physical cosmos to science. The effects of the Cartesian bargain continue to our modern times, when many scientists ignore and exclude immaterial and supernatural causes, and awkwardly fumble with materialist explanations for immaterial concepts like the mind, consciousness, and the origin of life and the cosmos.

Did the Cartesian bargain represent the first step toward scientism, and toward the atheistic materialist worldview?

Yes. By banishing anything related to God, angels, souls, or spiritual concepts, scientific exploration is blinded and silenced from exploring fundamental aspects of the human experience. Theology, spiritual understanding of the mind of God, and the belief in the rational ability of mankind to grasp the language and handiwork of God led to the Scientific Revolution. Yet the Cartesian bargain alienated from science the theistic ideas and concepts that delivered the scientific method.

Was Descartes an atheist or a believer in God? Did the philosophy of Cartesian dualism between the immaterial and material affect his religious beliefs?

Rene Descartes was a devout Roman Catholic Christian for his entire life. Cartesian ideas about the soul and the material world were earnest philosophical speculation, not a purposeful attempt to undermine theology or Christianity.

Yet impoverishing the systematic exploration of the natural world (science) by excluding from study God, angels, souls, and spiritualism had lasting effects. In this way, atheistic materialism traces a direct lineage to the Cartesian bargain in the 17th century. Ironically, Rene Descartes himself was not an atheist, nor a materialist, but the lines drawn from his radically new philosophy and metaphysics laid the groundwork for the scientism that emerged much later. Under scientism, atheistic materialism reaches its final stage of insularity, banishing any truth or claim to knowledge that cannot be physically established using scientific experiments under a materialist worldview.

What philosophical and metaphysical developments after the Cartesian bargain led to atheistic materialism and scientism?

In the 19th century, the mechanistic revolution banished any serious discussion of the topics that the Cartesian bargain relegated to theology or philosophy, like the soul, God, and spirits. After the Cartesian bargain isolated into the immaterial realm all the spiritual concepts (like God, souls, spirits, and angels), later skeptics, empiricists, and atheists banished them entirely from the academic and scientific discussion.

Our universe was reduced to a machine-like system of physical matter, operating under regular laws that could be understood and predicted like clockwork. This materialist reductivism gave no quarter for immaterial concepts like God, souls, spirits, and angels. Rather, any remaining belief in God was constricted into deistic thinking, whereby God is inactive after setting into motion the cosmos.

What is deism?

Deism holds that after creating the universe according to rationally ordered principles and laws that mankind can discover, God receded from any intervention or interaction with the cosmos. The universe was prepared by God from the beginning to operate on its own, independently from God.

What is the deistic worldview?

Under the deistic worldview, miracles are ruled out, and God does not intervene in the activity of the natural world but only set it in motion at the beginning. For a crude analogy of deism, God designed and manufactured a clockwork universe that runs on its own, and then God pressed the "start" button and retired from interaction with our existence and cosmos.

How does the deistic worldview fit well with atheistic materialism and methodological naturalism?

Atheism denies the existence of God; deism denies the activity of God in nature. Materialism precludes supernatural or divine causes; deism precludes the detection of continuing theistic intervention in nature. Methodological naturalism excludes any immaterial causes from investigation or consideration; deism professes that God excludes Himself from cause and effect in nature after setting everything in motion at the beginning of time. In this way, deism allows for the belief in God, including the Creation of the cosmos, but conveniently excludes any continuing influence or activity of God. For these reasons, the practical result of deism in scientific thinking is closer to atheism than theism.

15 Quantum Mechanics and Idealism

The principles of quantum physics are shockingly different from classical physics. While larger, readily observable matter behaves in predictable, mechanical patterns, the laws of physics change dramatically when examining the tiniest building blocks of matter itself. Matter, in the behavior and form of its smallest sub-atomic particles, can be described only using mathematical, probabilistic equations that only minds can understand, which lends credibility to idealism as the explanation for the nature of all reality. The indeterminacy of electrons, operating in a cloudlike region of possibilities, recalls the immaterial concept of potentiality and actuality first established by Aristotle, a connection noted by the great quantum physicist, Werner Heisenberg.

What is quantum physics or quantum mechanics?

Quantum physics reveals the mechanics, laws, forces, and physical behavior of the tiniest particles of matter called quantum particles. Quantum mechanics is distinguished from classical, Newtonian physics that simply does not apply at the quantum level of scale. Strangely, we must use two different sets of physics for the tiniest quantum particles (quantum mechanics) versus everything else in the cosmos (classical, Newtonian physics). The rules of quantum physics are entirely different, immensely complex, non-intuitive, and philosophically intriguing.

How does quantum physics differ from classical physics?

While classical, Newtonian physics would expect a linear, defined, predictable path—just like a planetary revolution or a thrown baseball—quantum mechanics is more probabilistic. You can predict exactly where the earth and moon will be, relative to the sun, depending on the time. But you cannot predict exactly where an electron will be, relative to the nucleus of an atom. The best approximations in quantum mechanics involve probability and wave functions, not any definite position within space-time. This uncertainty principle contrasts the probabilistic nature of quantum mechanics when compared to deterministic, classical, Newtonian physics.

Do the principles of quantum mechanics shed light on the timeless debate of idealism versus materialism? How does quantum physics demonstrate that at the greatest level of detail, even matter itself points to immaterial, rational minds?

Idealism holds that reality is mindlike at its core; materialism reduces everything to physical matter. While classical, Newtonian physics conforms easily to a materialist viewpoint, quantum physics—the behavior of the fundamental sub-atomic particles of all matter—supports idealism because quantum indeterminacy reveals probabilistic, discontinuous clouds of possibility that can only be described with the abstract language of mathematics.

Of course, mathematics is a rational language intelligible only to minds. Therefore, quantum physics demonstrates that even matter itself points to immaterial, rational minds. If the deepest scientific exploration of matter itself reveals mathematical, rational expressions rather than concrete, stable, predictable matter, then idealism has gained the upper hand over materialism.

Do quantum particles reveal predictable, linear patterns like classical, Newtonian physics? Or instead do quantum particles reveal indeterminate, probabilistic potentiality only expressible by mathematical wave functions?

Unlike larger objects subject to classical, Newtonian physics, the present or future state of quantum particles like electrons cannot be predicted or known. Quantum particles like electrons do not follow predictable, linear paths. Only when electrons are observed can we assign them a specific point in space and time. Before and after observation, sub-atomic electrons occupy a cloudlike, probabilistic potentiality, rather than any fixed position or predictable path of motion. In this way, quantum physics is inherently indeterminate and explores potentiality as much as reality.

What did the pioneer of quantum mechanics, Werner Heisenberg, notice about the mathematical, intangible, indeterminate nature of quantum mechanics and how it relates to the immaterial mind?

German theoretical physicist, Nobel laureate, and pioneer of quantum mechanics Werner Heisenberg (1901 - 1976) explained in The Physicist's Conception of Nature (1958) that particles are accessible only through abstract descriptions using numbers, equations, and mathematics, which sounds very much like the worldview of idealism:

"We can no longer speak of the behaviour of the particle independently of the process of observation. As a final consequence, the natural laws formulated mathematically in quantum theory no longer deal with the elementary particles themselves but with our knowledge of them. Nor is it any longer possible to ask whether or not these particles exist in space and time objectively. ... When we speak of the picture of nature in the exact science of our age, we do not mean a picture of nature so much as a *picture of our relationships with nature*. ... Science no longer confronts nature as an objective observer, but sees itself as an actor in this interplay between man and nature. The scientific method of analysing, explaining and classifying has become conscious of its limitations, which arise out of the fact that by its intervention science alters and refashions the object of investigation. In other words, method and object can no longer be separated."

16 Solid Objects Are Not Actually Solid, Undermining the Materialist Worldview

The materialist worldview is thrown into chaos by the discovery of quantum mechanics. Solid objects are anything but solid. Subatomic particles are humming with cloudlike patterns of activity only described through immaterial mathematical equations. Before observation, electrons occupy a cloudlike region of possibilities – in some ways, electrons are everywhere and nowhere. Strangely, electrons react to human observation by a "collapse of the wave function," when this cloudlike region of possibilities collapses into a discrete, fixed position. Materialism should expect simplistic models for particles like tiny billiard balls crashing about according to fixed, predictable laws. Yet quantum physics reveals an entirely different world in which matter itself is in a state of indeterminacy and constant motion, expressible only through immaterial equations intelligible only to immaterial minds.

Are solid objects – like bars of pure gold or bullets of ammunition – actually solid?

No. Material objects are not solid in the way most people understand the concept of solidity. While the gold bars will feel dense in your hands, and the bullets will shatter your bones if propelled fast enough, when you examine any material object at the sub-atomic level, the composition is strikingly different from your expectations.

What do the tiniest building blocks of material objects look like?

If you could magnify your vision of an object, using a magnifying glass or powerful microscope, that object will appear quite different from what you perceive using the naked eye. You may notice finer details and tiny structures invisible to the naked eye. If your microscope is powerful enough, you can examine the very building blocks of matter: molecules, which consist of two or more atoms, chemically combined.

What is a molecule? What is an atom?

Molecules are the tiny building blocks of matter, and molecules are made up of even tinier atoms.

The atom is the smallest unit of a chemical element, such as one unit (atom) of hydrogen, oxygen, helium, or carbon. If two hydrogen atoms are chemically combined with one oxygen atom, this fusion of atoms forms a single molecule of water. Chemists describe water using the familiar characters "H2O" to signify two hydrogen atoms combined with one oxygen atom.

What does it mean to examine something at the sub-atomic level? What is the quantum scale?

Examining things at the sub-atomic level means exploring the composition of the atom itself. Physical laws are strikingly different at the sub-atomic level, and these unusual laws of nature are called quantum mechanics or quantum theory to distinguish from the classical, Newtonian rules of physics that apply to larger objects. Magnifying your perception to the quantum scale of atoms concentrates your view to focus on yet smaller components – the sub-atomic particles: protons, neutrons, electrons.

What are the sub-atomic particles? What are protons, neutrons, and electrons?

Each atom has a nucleus at its center, consisting of protons and neutrons. Electrons travel in an orbit around the nucleus, spinning around and revolving, sort of like how planets revolve around the sun. Yet this metaphor of orbital revolution is imperfect and inaccurate because electrons and sub-atomic particles behave in strange ways through quantum mechanics.

Do electrons truly orbit around the atomic nucleus, like planets around the sun?

No. Electrons behave according to quantum mechanics, which differs from classical, Newtonian mechanics. The planetary metaphor is not perfect, and in some ways misleading, because electrons do not follow definite, linear paths around the nucleus.

Following quantum mechanics, the electrons occupy a cloudlike region of space around the nucleus. All we can do is calculate the probability of what space the electron may occupy at any given time, but we cannot know for certain. This probability distribution for all possible states of the electron is called the standing wave or the wave function. Using mathematics, we can calculate the frequency, wavelength, and speed of the wave function of the electron, but we cannot "know" the definite position or location of the electron without "collapsing the wave." Once we observe the definite position of the electron, the indeterminate wave behavior abruptly collapses into a fixed point.

What is the nature of a wave pattern? Do waves imply motion or stillness?

Waves imply motion, not stillness. Just like when you feel the effects of an ocean wave when riding in a boat, wave patterns occur over time. A wave is not a point in time; a wave is not still or fixed. The wave pattern is a repeating activity, and the movement takes time to occur.

Before and after observation, the behavior of electrons is wavelike, following the pattern of a mathematical wave function. Of course, waves "take time" to manifest, because waves represent movement over time according to certain parameters like frequency, speed, and amplitude. A mere moment within a wave is not truly a wave, but an incomplete strand thereof—merely a point in time.

What does quantum mechanics mean by the "collapse of the wave function" when observing electrons? How does the behavior of electrons change, once observed?

Instead of fixed positions, or defined paths, electrons occupy a cloudlike sum of probabilistic possibilities expressed as a mathematical wave function—until they are observed, which mysteriously solidifies their position in time and space.

Once we observe the actual position of the electron, the electron no longer exhibits wavelike behavior. Instead, the electron is observed to be stationary, in one position, hence the phrase, "collapse of the wave function." This confusing and mysterious aspect of quantum mechanics is one reason why it is so difficult to assign or detect or predict a specific location for electrons until actively observed.

Described as the "collapse of the wave function," the active observation of electron activity seems to constrain or "collapse" a multitude of mathematical potentialities into a recordable, finite, momentary existence. The fact that quantum mechanics and the behavior of the smallest units of detectable matter can be described only in the language of mathematics—using probability equations and wave functions—suggests an inherent and fundamental connection between matter and immaterial minds.

Why should probabilistic, mathematical, and wavelike behavior of sub-atomic particles point to immaterial minds rather than to matter?

Probability, equations, and wave functions are inherently generalized, abstract, rational, and immaterial. Mathematics is an abstract language that follows rational rules and can be grasped only by rational minds. Only an intelligent, immaterial mind can understand immaterial mathematics.

By magnifying all matter to the finest detail at the level of quantum physics, we find that the best scientific descriptions are encoded within probabilities and potentialities accessible only through the rational, abstract language of mathematics. The connection to idealism, and the departure from strict materialism, could not be more striking. While materialism predicts brute, concrete matter, idealism instead predicts an intelligent world of immaterial ideas, subject to rational laws and equations, only accessible by other intelligent minds. The evidence from quantum physics pulls the rug out from underneath materialism and rolls out a welcome mat for idealism.

Is our perception that material objects are stable and immobile an illusion? Is any material object truly at rest? Are inanimate objects continuously engaged in motion, change, and activity at the quantum scale?

All material objects consist of atoms at the smallest level of examination. These atoms consist of sub-atomic particles that follow quantum laws of physics, including the pattern of wave activity exhibited by electrons. Only through direct observation do any electrons occupy a fixed position, otherwise, they represent a probability cloud across space and time.

Therefore, even an object you consider to be "still," "fixed," and "not moving" is buzzing with (wavelike, indeterminate) motion, change, and activity at the quantum level. Our perception that objects have a concrete, stable nature is an illusion. The wavelike pattern of activity of the smallest bits of matter proves that no object is ever truly at rest, and that motion, change, and activity reverberate throughout every particle in the cosmos.

What does the most detailed examination of matter reveal about the debate between materialism and idealism? Does quantum physics undermine the materialist worldview?

Quantum physics reveals immaterial equations and mathematical wave functions, rather than discrete positions in space-time. Peering into the most detailed view of matter itself yields an abstract language, following rational laws and equations, intelligible only to immaterial minds. Perhaps the idealist worldview—holding that the ultimate reality is based on immaterial minds, not matter—is better supported by these startling revelations of quantum mechanics.

17 Why Does Mathematics Correspond to Reality?

Mathematics accurately and precisely describes the material world, but nobody questions how and why this is possible. The immaterial, rationalistic realm of mathematics has no physical or material connection to our physical, material cosmos – yet mathematics "works." Whether you believe that the language of mathematics was invented or discovered, the origin and nature of this language is entirely immaterial, so we should wonder how and why it corresponds to physical reality. Why can we understand messages written in the language of mathematics, and who is the messenger?

Was mathematics invented or discovered?

No definitive agreement exists about whether mathematics was invented or discovered. If mathematics is inherently derived from reality, mathematics has always existed, waiting to be discovered. If mathematics is instead viewed as an artificial tool and technique used by humanity at whim, mathematics is more like an invention.

What makes mathematics seem like it was invented?

The pioneers of mathematics like Archimedes, Pythagoras, Isaac Newton, Carl Friedrich Gauss, Leonhard Euler, Fibonacci, and Hypatia did not simply unearth the prize when digging with a shovel in the dirt. Nor did these intellectual titans look through a telescope to "find" mathematics. Instead, pioneering mathematicians applied abstract reasoning and logic to develop numerical representations and equations of the external world and of ideas internal within their immaterial minds. Without the intellectual work of geniuses, mathematics may never have become available. For these reasons, you could argue that mathematics was invented, not discovered.

What makes mathematics seem like it was discovered?

Mathematical truths seem to exist independently in the way that Greek philosopher Plato (427 - 347 BC) described. In the abstract world of Platonic Forms, the ideal form of the number five is eternal and cannot be destroyed or eliminated. We can count many tangible examples of the number five within our cosmos, but the immaterial idea or essence of the number five exists forever, untouched and undisturbed by our thoughts and meddling.

Mathematical laws and equations also correspond to patterns and behavior in nature. Snowflakes and flower petals follow geometric patterns of symmetry that can be expressed mathematically. The Fibonacci sequence and Golden Ratio were discovered throughout many natural phenomena: plant branches, seashells, pinecones, even the family pedigree of honeybees. Isaac Newton's (1643 - 1727) mathematical modeling of physics matches the behavior of the natural world with astonishing precision. Therefore, mathematics may have been revealed or discovered as an integral part of our cosmos, rather than artificially invented.

If mathematics was invented, why does it correspond to reality? If mathematics was discovered, why does an abstract, rational language, intelligible to immaterial minds, correspond to physical reality?

Both questions cut to the heart of the mystery of mathematics and our cosmos. Familiarity numbs us to the shocking and dazzling correspondence of the physical world with the abstract, rational language of numbers. The immaterial idea or essence of the number five does not exist in nature, but only within our immaterial minds (and perhaps in other realities, according to idealists like Plato or Pythagoras). Yet the abstract, immaterial idea of the number five and all its friends are incredibly useful allies in our efforts to describe, map, predict, manipulate, and understand the physical, tangible, material cosmos.

What is so intriguing about the connection between mathematics and the physical world? What are the implications of mathematics being invented versus discovered?

No matter whether mathematics was invented or discovered, we should be surprised and intrigued that the intangible language of numbers matches the tangible universe. If mathematics was invented, we should be shocked that humanity devised a self-enclosed, independent, abstract, rational language or system that matches the physical behavior of the cosmos. Why should our invented techniques work so seamlessly to describe and predict the universe and its laws of motion?

If mathematics was discovered, the provocative idea is that reality itself is translatable and intelligible using abstract, rational language intelligible only to immaterial minds. If there is a language of the cosmos, and we know how to speak and understand this language, who is speaking? Who or what designed this language? Why can we understand messages using this language? What is the message and who is the messenger?

18 Laws of Nature Do Not Cause Anything

Humans invented the concept of "laws" of nature to correspond to legal systems. Laws must be followed. But while humans can violate criminal law, the laws of physics are inviolate, consistent, predictable, and reliable. We live in an ordered, not chaotic, cosmos. What mysteries are unraveled by the order, rationality, and fine-tuned structure of the cosmos, and by the intelligibility of the laws of nature by human minds?

What compels matter to follow the laws of nature?

Materialism has no answer for why matter follows the laws of nature. While physics offers a multitude of equations, ratios, forces, and so-called laws of nature, none of these parameters explain the behavior they describe. Science (and physics and mathematics) can only describe the movement of matter.

How did C.S. Lewis illustrate the difference between descriptive science and the underlying reason or explanation for what science describes?

British scholar and theologian C.S. Lewis (1898 - 1963) humorously quipped:

"To say that a stone falls because it is obeying a law makes it a man and even a citizen."

Lewis thereby illustrates the humanity or agency we unconsciously inject into material objects when describing their behavior using scientific language. Yet only living creatures have the agency to make choices or enforce their will on other objects. Matter makes no choices and technically does not "follow" or "obey" laws of any type.

By our human-centered metaphors, we unconsciously (and falsely) imply that material objects do things, follow laws, change directions, reverse polarity, split into two waves, etc. but these descriptions are ultimately misleading. Matter has no choice, no will, no conscious activity. Matter is acted upon, and we describe the results using laws, equations, and other descriptions.

Do any laws or equations of motion, or fundamental forces like electromagnetism and gravity, tell us why things behave as they do?

No. Physical laws, equations, and forces tell us nothing about why these bodies or particles behave as they do. What mathematical modeling of physics tells us is how things behave in a descriptive manner. Mathematical modeling—equations, proportions, inverse relationships, laws, constants—is a descriptive process answering the question of "how" but not "why." For example, we can measure the force of gravity but not explain why gravity attracts matter together.

What does it mean to question why the laws and equations of physics accurately and precisely describe the motion and behavior of our universe?

Mysteriously, immaterial minds somehow connect with the material cosmos. We take it for granted that physics equations and laws "work." By "work," we mean that they conform to the reality of our observations. We also mean that these mathematical relationships can predict behavior, thereby corresponding to future events.

By expressing the motion of objects in the language of mathematics, we can observe and predict the natural, concrete, tangible world. Yet mathematics is an entirely abstract, intangible, immaterial system. We should be surprised by and marvel at the fundamental connection between the abstract, immaterial world of mathematical equations and the concrete world of nature and moving objects.

Why should we be surprised by and marvel at the precise, accurate, and reliable match between abstract, intangible mathematical language and the concrete, tangible motion of objects in nature?

Abstract, intangible systems like mathematics point to the abstract, intangible world of the immaterial mind, thinking, rationality, and consciousness. Inanimate objects do not think, do not have minds, and have no consciousness nor rationality. Yet the behavior of tangible, concrete objects can be expressed precisely using intangible, abstract language. Further, we should marvel at the order, uniformity, and reliability of mathematical descriptions for the behavior of material objects. Put simply, "they work!" but we never stop to consider how chaotic and unpredictable our cosmos could be if the "laws" were instead observed as mere "suggestions."

Therefore, we should be surprised at the sheer improbability that immaterial mathematical language corresponds so precisely to the behavior of our material cosmos. We should marvel at this fact because it suggests there is immaterial mind, thought, and consciousness at the source of all material, including the physical laws, equations, and fundamental force of the cosmos. The order of our cosmos is the handiwork of design, and all designs have a designer.

What did Albert Einstein say about the comprehensibility of our sensory experience through science, mathematics, reasoning, and physics?

Theoretical physicist Albert Einstein (1879 - 1959) wrote the following in a 1936 article:

"The eternal mystery of the world is its comprehensibility…The fact that it is comprehensible is a miracle."

19 The Faustian Bargain: Theocentric vs. Anthropocentric Worldviews

The struggle between the worldview of God as Creator versus man as creator is dramatically told and retold in countless literary and artistic works, including through classic characters like Prometheus, Faust, and Dr. Frankenstein. Whether God or mankind is at the center has profound effects on your worldview, identity, and everyday behavior. By examining a few of the most famous cultural examples that contrast the theocentric (God-centered) versus anthropocentric (mankind-centered) worldviews, we reveal how deeply this question is buried within the human psyche.

What is a theocentric or God-centered universe?

The theocentric universe has God as the source, the center, the ground of all being. God, the Creator, is the supreme organizer and arbiter of reality and truth. God created mankind. God created the universe. Mankind owes everything and all existence to God. Mankind is subject to the natural laws established by God.

What is the anthropocentric or human-centered universe?

The anthropocentric universe places mankind at the center of the cosmos. Mankind, the mind of man, becomes the supreme organizer and arbiter of reality and truth. God is discarded, ignored, rejected. Mankind owes nothing to God and is subject to no laws other than those self-imposed by individual or collective consent.

What did Blaise Pascal say about the longing for God within every human? What is the shortened, paraphrased version of Pascal's quotation?

Blaise Pascal (1623 - 1662) explained that human happiness can be fulfilled only through full communion with God:

"What else does this craving, and this helplessness, proclaim but that there was once in man a true happiness, of which all that now remains is the empty print and trace? This he tries in vain to fill with everything around him, seeking in things that are not there the help he cannot find in those that are, though none can help, since this infinite abyss can be filled only with an infinite and immutable object; in other words by God himself."

You may have encountered this quotation shortened and paraphrased as:

"There is a God-shaped vacuum in the heart of every man, and only God can fill it."

Who is the god of atheism?

The self. You. The search and longing for a higher power, a transcendental creator, a divine source of all being is related to the same human need to find meaning, purpose, and value. Those who reject God, no matter how forcefully they protest, cannot suppress their fundamentally human drive for worship, communion, and integration with the divine. By denying God, the atheist turns this healthy human appetite for purpose and meaning inward, into an unhealthy habit, obsession, and character trait of self-worship. The god of the atheist is himself: self-worship, the original sin of pride.

How did C.S. Lewis describe the inversion of the relationship between God and mankind?

In his book, The Abolition of Man (1943), British scholar and theologian C.S. Lewis (1898 - 1963) wrote:

"For wise men of old the cardinal problem had been how to conform the soul to reality, and the solution had been knowledge, self-discipline, and virtue. For magic and applied science alike the problem is how to subdue reality to the wishes of men."

What mythical stories and literature demonstrate the tension of human technology and ambition wrestling between the contrasting worldviews of the theocentric versus anthropocentric universe? How does the theme of "Creator versus creature" show up persistently in human art, literature, and storytelling?

The ancient Greek myth of Prometheus, Goethe's 18th-century play, "Faust: a Tragedy," and Mary Shelley's Gothic horror novel, Frankenstein (1818) all demonstrate the tension between the role of Creator (God) versus creature (mankind), as well as the role of technology in navigating the balance between the mortal and the divine.

What is the ancient Greek myth of Prometheus? How does Prometheus represent the theological theme of Creator versus creature?

Prometheus defied the ancient Greek gods by stealing fire to give to mankind. In the ancient Greek myth of Prometheus, the fire represents knowledge and technology. By stealing from the gods the creative power of fire, technology, and knowledge, Prometheus tried to reverse the balance of power between gods and men. Prometheus arranged by force an anthropocentric revolution, overthrowing the theocentric rule of the Greek pantheon of gods. The myth of Prometheus also exalts science and technology as divinely inspired and therefore a method for man to achieve the power and greatness of the gods.

What is the story of the play called "Faust: A Tragedy" written by Goethe? How does the story of the protagonist Faust represent the theological theme of Creator versus creature? What is the Faustian bargain?

Johann Wolfgang von Goethe (1749 - 1832) wrote the famous play "Faust: A Tragedy" about an ambitious academic who signs in his own blood a contract with the devil. Not content with what God provides and permits, Faust trades his soul to the devil—the adversary of God and humanity—in exchange for knowledge, power, and sensual pleasure. Exploring themes of good and evil, temptation and salvation, and especially human hubris, the "Faustian bargain" (with the devil) is forever cemented in the literary and artistic canon.

Whenever humanity strives to usurp the role of God as Creator, or makes ethical concessions to achieve worldly, scientific, or technological power, the familiar reference is to the Faustian bargain from Goethe's famous play. The Faustian bargain is a constant reminder of the great cost of grasping for an anthropocentric worldview, rather than accepting mankind's gifted grace within a theocentric reality.

What is the plotline of the 19th-century novel, Frankenstein? What is the full title of Frankenstein by Mary Shelley? How does the story of Dr. Frankenstein's monster explore the theological themes of Creator and creature?

Mary Shelley (1797 - 1851) wrote the novel Frankenstein in 1818. What few realize is that Mary Shelley wrote not just a horror novel, but one of the first examples of science fiction. The full title of the Gothic novel is Frankenstein; or, The Modern Prometheus, a direct reference to the Greek myth of stealing fire or technology from the gods to give greater power to mankind.

The mad Doctor Frankenstein somehow breathes life into stitched-up remains of body parts robbed from corpses and graveyards. Through mysterious technology harnessing the electrical power of the "spark" of life, Dr. Frankenstein's monster comes to life. Using science and technology, mankind rather than God has created a disgusting, freakish monster who struggles to live peacefully in a world that hates and fears the outsider. Dr. Frankenstein's monster kills the wife and also the best friend of his mortal creator before Dr. Frankenstein dies of exhaustion while seeking revenge against his created monster. Like the downfall of Goethe's Faust, Dr. Frankenstein demonstrates the hubris of mankind striving to usurp the role of God as Creator.

What 21st-century technology trend best exemplifies the Faustian bargain and the hubristic desire to usurp the role of God? What is the fantasy of Frankenstein's monster updated for the 21st century?

The emergence of new life in the form of strong or general artificial intelligence is the fantasy of Frankenstein's monster recurring in the 21st century. Breathless marketers of software startups assure the investing public that science fiction will merge with reality. Delusional technocrats, operating from a misguided, atheistic materialist worldview, believe that computer programs will give birth to a new life form called artificial intelligence. The ghost will emerge from the machine, they say. Any day now!

Why is the empty promise of artificial life emerging from artificial intelligence software the 21st-century version of Mary Shelley's Frankenstein? What 21st-century technology represents the Promethean fire stolen from the gods?

At the turn of the 19th century, Mary Shelley wrote the timeless science fiction and horror novel about Dr. Frankenstein's monster. During Shelley's lifetime, the most mysterious and exciting technological frontiers concerned electricity. Just a few years after Shelley's 1818 novel, Michael Faraday invented the electric motor. Mary Shelley's circle of friends included intellectuals and scientists who discussed the divide between life and death, and whether the "spark" of life may come from electricity, as suggested by grisly demonstrations of electrical jolts applied to dissected frogs and even human cadavers. Naturally, electricity played the part of the fire stolen from the gods to give life to long-dead body parts.

At the turn of the 21st century, computer software—in the overhyped and misleading form of so-called "artificial intelligence"—represents the fire stolen from the gods to elevate mankind to the divine throne. Instead of Dr. Frankenstein's electricity animating a composite heap of cadaver limbs, now it is software that will animate a sequence of software code. The ghost will emerge from the machine, say the marketers and entrepreneurs of Silicon Valley.

The promise of life emerging from software is the same hubristic fantasy displayed in Mary Shelley's 19th-century blend of horror and science fiction. Yet in the 21st century, technocrats who deny God and assert that everything is material profess on blind faith that man will create life using computer programming language. Supposedly, the divine spark of life will emerge—any day now!—once the software code reaches the required level of complexity. For some reason, the prophets of artificial life keep pushing back the date of this electronic apotheosis, no doubt to keep up with quarterly earnings reports.

What do atheistic materialist technocrats call the prophesized moment when artificial life emerges from computers?

Ironically, technocrats call the emergence of new life from software code "the Singularity." The irony is that the Singularity is the same alternate name for the mysterious Big Bang, when all of time, space, matter, and energy appeared to form the beginning of our entire universe. Now when someone asks you to distinguish which Singularity you mean, you have the choice of pointing to the beginning of all Creation or the technocratic promise of the day when computers became alive.

If you have noticed the distinct signs of an atheistic religion for materialist techno-junkies, you are paying attention. This second Singularity or technological creation event hopes to eclipse the Creation of our cosmos and everything in it, but this time, mankind wriggles his way onto the grand stage to steal the honors.

What is hubris?

One definition of hubris is excessive pride or self-confidence leading to downfall.

Which of the Seven Deadly Sins is closest to hubris?

Among the Seven Deadly Sins identified by Christian theology, pride is most closely related to hubris. The sin of pride refers to the unhealthy love of self or the desire to be better than others. Christians value humility rather than pride as exemplified by the self-sacrificial love of Jesus Christ, and by the Incarnation of God as a human being, walking in the same footsteps as mankind.

Why is pride the most dangerous of the Seven Deadly Sins?

All other sins prey upon the weakness of human nature. Gluttony preys upon our need for food and nourishment. Lust exploits our desire for love and procreation. But pride is a sin of the spirit or soul. As British scholar and theologian C.S. Lewis (1898 - 1963) explains in his 1952 book, Mere Christianity, chapter 8, "The Great Sin":

"Pride gets no pleasure out of having something, only out of having more of it than the next man. We say that people are proud of being rich, or clever, or good-looking, but they are not. They are proud of being richer, or cleverer, or better-looking than others. If someone else became equally rich, or clever, or good-looking there would be nothing to be proud about. It is the comparison that makes you proud: the pleasure of being above the rest. Once the element of competition has gone, pride has gone. That is why I say that Pride is essentially competitive in a way the other vices are not."

"The Christians are right: it is Pride which has been the chief cause of misery in every nation and every family since the world began. Other vices may sometimes bring people together: you may find good fellowship and jokes and friendliness among drunken people or unchaste people. But Pride always means enmity—it is enmity. And not only enmity between man and man, but enmity to God.

In God you come up against something which is in every respect immeasurably superior to yourself. Unless you know God as that—and, therefore, know yourself as nothing in comparison—you do not know God at all. As long as you are proud you cannot know God. A proud man is always looking down on things and people: and, of course, as long as you are looking down, you cannot see something that is above you."

What is the cosmic irony of atheistic materialists believing life will emerge from machines?

The cosmic irony is that atheistic materialists deny that humanity has a soul, deny purpose in the expansion of diversity of life on earth, reduce the mind to a lump of brain tissue, and grope for impossible explanations of infinite universes to explain away the stunning improbability of all matter, energy, time, and space appearing in the Creation event that scientists call the Big Bang. All these timeless mysteries of the mind, the soul, and the Creation of our cosmos are brushed off as superstitious beliefs. Yet the same technocrats who reject God place their faith in themselves as the false god of an imaginary computer lifeform. Prometheus has stolen from the gods again, but he does not yet realize the goods are counterfeit.

Part 3, Mind and Consciousness

20 Introduction to the Immaterial Nature of the Mind

Materialism cannot account for the mind because minds are immaterial. The "hard problem" of explaining how our immaterial consciousness emerges from brute matter is a challenge only to atheistic materialists because their worldview limits all explanations to matter in the form of molecules and atoms. Ironically, the careful study of atomic particles through quantum physics cries out for non-material, mindlike explanations for the vast amounts of information therein, expressible only in the immaterial, abstract, rational language of mathematics that is intelligible only to immaterial minds.

How depressing, impoverished, and limited is the atheistic materialist worldview that all reality, all truth, all beauty, all goodness is a delusion or nothing more than unconscious matter, crashing about by blind, random forces. Your choice to believe that mind precedes matter, or that matter precedes mind, explores the heart of existence itself, and pulls your inner heart along for the journey.

Reducing your immaterial mind—your power of free will, intellect, rational thought, and abstract reasoning—to an illusory product of the lump of soft tissue encased in your skull is an intellectual, spiritual, and humanitarian abomination. Our immaterial minds are not limited to the materialist worldview, nor are they limited to brain function.

Even the sober scientific study of the mind-brain connection gives distinct clues that tear through the gauzy screen of the materialist dogma. Minds can function without the brain, and the connections between activity of the mind and brain have not been completely explained, despite exhaustive and painstaking scientific exploration. Most significantly, of those precious things that cannot be reduced to matter and material causation, minds join the glorious list, along with love, beauty, goodness, truth, and their ultimate source—God.

21 Free Will

Some atheistic materialists are so frustrated by the inability of their worldview to explain free will that instead of changing their belief system, they deny the existence of free will. Strange philosophical decisions (like denial of free will) highlight catastrophic problems within the atheistic materialist worldview that cannot account for the immaterial mind, consciousness, and free will experienced by every person.

What is free will?

Free will is the ability to choose your actions through conscious deliberation. Even if you react to emotions, impulses, desires, or temptations, your free will controls your final decision. Despite an intense desire to eat candy, or temptation to lie, or anger that prompts you to scream, you can exercise your free will to resist these impulses.

Does anyone deny that human beings have free will?

Surprisingly, some atheistic materialists deny that free will exists. Instead, some believe that everything we do is determined by external pressures and natural events. If the universe consists only of matter, with no purpose, no meaning, and no consciousness, then our actions are determined entirely by our environment, and by the brain tissue in our skulls. Atheistic materialism denies the existence of a soul or spirit, so, acknowledging our free will may creep uncomfortably closer to forbidden territory.

What are some dangerous implications of the denial of free will?

Denying free will diminishes humanity, removing a moral and spiritual obstacle to abuse. Viewing people as unthinking animals, brutishly responding to stimulus, can lead to tremendous evil like oppression or mass murder. The social scientist Hannah Arendt (1906 - 1975) identified dehumanization as the critical first step leading to the brutal crimes and genocides committed by totalitarian governments.

Further, denial of free will precludes moral responsibility and legal culpability. If we make no choices, we cannot be responsible for our choices. Crime, sin, right, and wrong become irrelevant if there is no free will. Our society requires objective moral responsibility, and our legal system is based on culpability, guilt, and punishment for wrongful behavior. With no free will, there is no basis for societal morality or systems of crime and punishment.

Is the denial of free will inherently self-refuting?

Yes. You should not take seriously the argument that free will does not exist because the argument is self-refuting. Those who deny free will are stating (as a result) that they cannot make conscious decisions or think for themselves. Their words and actions are mere reactions to physical stimulus in the universe, like crumbling rocks tumbling down a mountainside. Therefore, anything they say has no more meaning, relevance, or truth than the sound of wind blowing through the trees.

Is the denial of free will consistent with human behavior?

No. The denial of free will is inconsistent with human behavior. Everyone who tries to deny free will exposes the absurdity of the idea by their own behavior. Observe anyone, even one silly enough to deny free will, and you observe someone who lives as if free will is a reality: asking and granting permission, making promises, contemplating decisions, and blaming others for crimes and misdeeds.

What is one possible motivation for the extreme claim that free will does not exist?

Scratch a free will denier and you will smell an atheistic materialist. The denial of free will evokes the familiar saying that for some people, "There are three truths: what you tell yourself, what you tell others, and what you truly believe." Nobody actually believes that free will does not exist, or rather, nobody acts as if it were true.

Denial of free will is a dramatic signal of dedication to atheistic materialism. While nobody lives life as if free will is fiction, making the radical claim is a dramatic endorsement of "faith" in the worldview that everything is material and therefore humans are also mere objects reacting to stimulus. Denying free will also dodges The Hard Problem of Consciousness, which cannot account for immaterial minds making conscious decisions if we are nothing but unconscious, material molecules.

22 The Hard Problem of Consciousness and Panpsychism

Atheistic materialism cannot account for consciousness because unconscious matter has no mechanistic connection to the immaterial mind. Rather than confronting The Hard Problem of Consciousness – explaining how immaterial consciousness emerges from brute matter in the brain – some atheistic materialists instead propose that consciousness is everywhere. Instead of explaining immaterial consciousness of the mind, the panpsychist worldview multiplies, expands, and infuses consciousness within every particle of matter in the universe.

Just like the preposterous attempt to explain our fine-tuned universe as one of trillions of alternate (yet undiscovered) universes, panpsychism explains consciousness by denying its rarity and imagining that consciousness is within every particle in the cosmos. If our universe is only one of an infinite ensemble, maybe it is not special. If our consciousness is simply aggregated from the consciousness of atoms and molecules, maybe we are not special. Both excuses from atheistic materialism about the fine-tuned cosmos and immaterial consciousness are fantastical, evidence-free, and against common sense within human experience. We do not encounter alternative universes or consciousness inside rocks. Instead, our minds, our consciousness, and our fine-tuned cosmos point to immaterial phenomena incompatible with atheistic materialism.

What is consciousness? What is the mind?

Your mind and consciousness are fundamental to your identity and your experience as a sentient, thinking person. As concepts to study, the mind and consciousness are closely related. The Merriam-Webster dictionary definition of consciousness references the mind: "the state of being characterized by sensation, emotion, volition, and thought: mind."

Merriam-Webster defines the mind as "the element[s] in an individual that feels, perceives, thinks, wills, and especially reasons." Both definitions focus on mental activities of human beings: perception, emotion, free will, abstract reasoning, and other thought processes. The same dictionary defines the verb "to think" as "to form or have in the mind." Now, where or what is the medium or location or process to form and hold these thoughts? Is it material or immaterial existence? How you think about where or how thoughts begin—or whether thoughts even exist— reveals important conclusions about your philosophical worldview.

Can you deny that consciousness exists? Is denial of consciousness self-refuting?

There is no rational way to deny that consciousness exists. If you deny that consciousness exists, then you have no consciousness. If you have no consciousness, there is no reason to consider anything you have to say or think.

Denial of consciousness is self-refuting and nonsensical. Hundreds of years ago, Rene Descartes (1596 - 1650) explained that by recognition of your own thoughts and consciousness, you have certainty of your own existence. Yet some dogmatic materialists reject the seemingly obvious and undeniable statement of Descartes, "I think, therefore I am." These supreme skeptics would admit that we exist, but counter that our consciousness is mere illusion, diminishing our existence to unconscious activity in a strictly material world.

Why would someone try to deny the existence of consciousness?

Denial of consciousness is yet another casualty of the denial of God and everything immaterial. The cynical reason to deny the existence of consciousness is to defend the atheistic materialist worldview, which denies the existence of anything other than material objects. Consciousness is immaterial and so it must be abandoned.

Materialism also believes that everything we experience happened by random, blind, unguided, material forces, which could not possibly create an immaterial, thinking, perceiving consciousness. As a result, the most committed and dogmatic atheistic materialists try to dismiss the existence of consciousness entirely. The desperate attempt to erase consciousness is a strategic maneuver to resolve one of the most glaring and widely recognized contradictions of the limited worldview of atheistic materialism, The Hard Problem of Consciousness.

What is The Hard Problem of Consciousness?

The Hard Problem of Consciousness is the glaring error or contradiction of the atheistic materialist worldview. If everything in the universe is exclusively material, including human beings, then how can immaterial consciousness emerge? How can immaterial consciousness emerge from unconscious matter, like the molecules and atoms that make up the brain?

For the materialist, what is the leading candidate for the physical source of consciousness of human beings?

Materialists assume that the brain is the physical source of consciousness. The organ found in the skull, although completely material, is somehow special or different from every other organ like the liver or heart. Although the mechanism is not explained, it is simply assumed that the brain produces or creates consciousness in some mysterious fashion.

How can a material organ, like the brain, produce or create consciousness?

This question restates The Hard Problem of Consciousness, which has not been solved, and remains a troubling mystery to atheistic materialism. Due to this obstacle, some atheistic materialists take the extreme and dubious position that consciousness does not exist at all, nor does the mind. Instead, only the material brain is real and your mind and consciousness are fantasies. This ridiculous worldview is called eliminative materialism because it eliminates the mind and consciousness.

Why is it called The Hard Problem of Consciousness? Who invented the term for the failure of materialism to account for the immaterial mind?

The Hard Problem of Consciousness earned its name because the problem remains unsolved, with no satisfying answer or resolution from atheistic materialism. One of the greatest challenges and criticisms of the materialist worldview is the existence of consciousness, something everyone experiences but nobody can explain under materialist assumptions. Therefore, The Hard Problem of Consciousness has earned its title. Australian philosopher David Chalmers (born 1966) introduced the term in his 1995 article, "Facing Up to the Problem of Consciousness."

What is another creative solution proposed by materialists to The Hard Problem of Consciousness?

Instead of denying consciousness entirely, or reducing it to brain function, some materialists propose that consciousness is everywhere, existing in every atom in the universe. Therefore, human consciousness is not special, nor a mystery, nor does it require any special explanation. The idea that consciousness is found within every particle, everywhere, is called panpsychism.

What does panpsychism mean, literally?

Panpsychism derives from the Greek words "pan" (all) and "psyche" (mind or soul). Therefore, panpsychism supposes that consciousness is found everywhere, within everything. Not just human beings but also animals, plants, and even inanimate objects like rocks and crystals contain consciousness. Even the microscopic building blocks of material contain consciousness, so molecules and atoms are conscious, also.

How does panpsychism provide a solution to The Hard Problem of Consciousness?

According to the theory of panpsychism, The Hard Problem of Consciousness disappears because there is no need to explain how unconscious brain matter produces consciousness. Instead, the brain tissue is already conscious because all matter is conscious. Everything in the universe is conscious, not just the brain tissue of human beings. Therefore, panpsychism trades the mystery of how the human being experiences consciousness for the sweeping assumption that the entirety of existence, all matter in the universe, is already conscious. If you find this "explanation" lazy, far-fetched, and suspiciously convenient, you are not alone.

Is there any similarity between the panpsychism "solution" for The Hard Problem of Consciousness and the Multiverse "solution" for the fine-tuned universe?

Yes. The similarity is that both "solutions" use grandiose and sweeping assumptions to spread and multiply the mystery to new frontiers. Human consciousness need not be explained because literally everything is conscious. The fine-tuned universe requires no explanation because of infinite alternate universes—collectively, the Multiverse (see chapter 12, "The Multiverse and Ockham's Razor"). If you examine these so-called "solutions" with a critical eye, two faults are obvious.

What are two of the largest problems with panpsychism and the Multiverse as convenient excuses for weaknesses in the atheistic materialist worldview?

First, these so-called "solutions" violate Ockham's razor, egregiously, by multiplying entities at grand scale. Instead of focusing on the origin of consciousness of human beings, panpsychism injects consciousness into every particle of matter in the universe. Instead of examining our known universe and its fine-tuning, the Multiverse imagines an infinite number of other fictional universes that nobody has ever experienced, detected, or proven.

Second, these so-called "solutions" are thin soup. If you add too much water to soup, the flavor and impact of the main ingredients is diluted, lost amidst a sea of flavorless broth. Similarly, by spreading consciousness throughout the universe, consciousness becomes common, pedestrian, universal—nothing special or unique.

For the Multiverse theory, the multiplication of infinite alternate universes means that our known universe is nothing special or unique, but simply one universe within the immense scale of the Multiverse. Both panpsychism and the Multiverse theory foolishly try to explain a single, defined mystery by multiplying that same mystery across the entire universe (panpsychism) or by multiplying entire realities (the Multiverse theory).

Instead of adding water to stretch the volume of the soup, thereby diluting the main ingredients, better explanations would carefully explain the main ingredients (consciousness, fine-tuning). Yet The Hard Problem of Consciousness and the fine-tuning of our cosmos point to immaterial concepts that contradict the atheistic materialist worldview. Instead of diligently investigating these clues, atheistic materialists further complicate the mystery.

23 Is Your Mind Nothing but the Product of Brain Tissue?

The mind exists beyond the brain, although the mind exercises the brain as a bodily organ involved in mental activity. Atheistic materialism denies the distinction and independence of the mind, arguing that anything mental is merely the product of the physical brain. Yet neuroscientific evidence reveals that mind-brain dualism is more persuasive than reductive, mechanistic, materialist models of the mind-brain connection.

Where is your mind, according to the atheistic materialist?

The atheistic materialist believes your mind, if the mind exists at all, is located exclusively within your skull, inside the brain. Atheism denies the existence of God and souls and spirit. Materialism denies anything immaterial like minds or consciousness, so these concepts must be reduced to mere matter, like brain tissue.

What is neuroscience?

Neuroscience is the study of the structure and function of the brain. Given the physical focus of the field, many but not all neuroscientists are materialists.

Within the field of neuroscience, what is an eliminative materialist? What is a functional materialist?

According to the functional materialists within neuroscience, your brain produces the thoughts and consciousness that we call the mind. So, the mind is merely the function of the brain, hence the term, functional materialist.

Incredibly, eliminative materialists go further and believe the mind does not exist at all. To the eliminative materialist, we have no mind or consciousness. Any contrary idea is simply "folk psychology," a term of derision like fictional "folk tales." Everything we experience of consciousness, thought, emotion, sensation, and memory is a total illusion that does not exist. The only "real" thing is the physical brain, which somehow tricks us all into believing we are more than a meat-machine.

Is the idea of eliminative materialism a result of inherent weaknesses or glaring contradictions within the materialist worldview?

Yes. Trying to convince others that the mind does not exist is a desperate attempt to salvage the shambles of the atheistic materialist worldview. Strict materialism cannot account for the immaterial mind nor explain the many lines of scientific evidence—covered in later chapters—that refute this reductive point of view.

Instead of acknowledging that not everything is material, and thereby abandoning the worldview of strict materialism, some choose to believe the mind does not exist. This extreme position is a philosophical white flag of surrender to the overwhelming evidence and universal experiences that prove the mind is real and a fundamental aspect of human nature. If atheistic materialism had an adequate explanation for the immaterial mind, there would be no need to deny its existence. Therefore, feeble attempts to eliminate the mind or reduce it to brain tissue only draw attention to the limits and consequences of adopting the flawed worldview of atheistic materialism.

24 The Mind-Brain Connection and Mind-Brain Dualism

Mind-brain dualism establishes that the immaterial mind is beyond and distinct from the material brain. Atheistic materialism denies mind-brain dualism, and instead proposes several reductive, mechanistic relationships between the mind and brain. Functional materialism, identity theory, and eliminative materialism are worldviews that grapple with The Hard Problem of Consciousness in different ways, but they all deny the immaterial mind and try to reduce our fundamental identity to physical matter – brain tissue encased in the skull.

To the atheistic materialist, can your mind exist without your brain?

No. To the atheistic materialist, if your brain is destroyed, your mind is destroyed. If your body dies, your brain tissue dies, your brain stops functioning, and your mind disappears.

Is there neuroscientific evidence that the mind persists without the brain or that the mind is distinct from the brain?

Yes. You can explore a wide variety of neuroscientific evidence for the mind beyond the brain in the next several chapters.

To the atheistic materialist, can your consciousness exist without your brain, or can your consciousness be extended beyond your brain?

No. To the atheistic materialist, your consciousness is produced by your brain, and any experience or perception that your consciousness is beyond the brain is an illusion or delusion. For example, despite thousands of documented cases, any out-of-body experience of perceiving, seeing, or hearing anything is strictly denied by the atheistic materialist worldview. See chapter 33, "Near-Death Experiences."

Has neuroscientific research found the part of the brain that produces consciousness?

No. The unsolved mystery of how consciousness emerges from unconscious matter is called The Hard Problem of Consciousness. In fact, in 1998, philosopher David Chalmers (who coined The Hard Problem in a 1995 essay) bet neuroscientist Christof Koch a case of fine wine that in 25 years, scientific efforts would not discover the physical mechanism or location for consciousness within the brain.

At the 2023 conference for the Association for the Scientific Study of Consciousness, Chalmers was declared the winner of the bet and Koch presented Chalmers with the wine. Neuroscience had not provided sufficient proof of its materialist assumption that consciousness is confined entirely to the physical brain.

To the atheistic materialist, can your consciousness, soul, or mind survive after death?

No. To the atheistic materialist, death marks the end of consciousness, with no exceptions. When you die that is the end of you. There is no life after death, nor does any shred of consciousness remain after death. Since consciousness is assumed to be a product of the brain, when the brain dies, consciousness dies with it. Atheists deny that the soul exists even during life.

What is the brain? How does the brain relate to thinking and the human mind?

The brain is an organ of the human body located within the skull. Unquestionably, the brain is involved in the function of thinking and cognition. The question is whether the brain is the exclusive source of our thoughts and mind, or rather an organ that assists or interacts with the thinking process and human mind.

Does the brain explain everything about the mind?

No. Yet some neuroscientists and philosophers, especially atheistic materialists, want to convince others that the brain explains everything about the mind. The worldview that the brain fully explains the mind is inherently promissory because it predicts discoveries and explanations that do not yet exist. Ironically, it is faith in the atheistic materialist worldview—not evidence—that drives the hope of solving the mysteries of the mind.

Materialist, reductive, mechanical worldviews deny any meaningful distinction between the mind and brain, believing that they are essentially, experientially, or functionally the same. Only because some aspects of the mind have been explained as related to the brain, atheistic materialists place their faith in the hope that all remaining mysteries will be solved, despite the contradictory or missing evidence.

What are some of the functions or activities of the mind (or brain)?

The mind (or brain) is involved in human perception, memory, emotion, motor skills, abstract thought, and free will (decision-making).

What are some divisions of belief within the materialist worldview of the mind-brain connection?

The reductionist or functionalist worldview of mind-brain materialism holds that any aspect of the mind, as a concept or activity, is nothing but brain activity. Therefore, the mind is reduced to the brain. In this worldview, without a brain, there is no mind to discuss or evaluate. The mind is what the brain "does" so the mind is merely the function of the brain.

The identity theory of mind-brain connection holds that the mind and brain are the same. In other words, the mind is simply a concept to describe our experience of using our brains. The mind is another name for the brain.

The eliminative worldview believes that the mind does not exist. Instead, the mind is a form of "folk psychology," or illusion, or misconception, concocted by those ignorant of the processes of the physical brain.

What is the mind-brain dualist worldview?

To the mind-brain dualist, there is a distinction between the immaterial mind and the material brain. Under mind-brain dualism, physical brain function is not sufficient to explain or generate all the activity, consciousness, and existence of the immaterial mind. Instead, the immaterial mind consists of something more than the activity of the physical brain tissue encased in the skull. Further, the mind—or at least substantial functions and activity of the mind—can and does persist in the total or partial absence of living brain tissue.

While mind-brain dualism does not deny the many connections between the mind and brain as also believed by materialism, dualism goes further. Mind-brain dualism proposes that there are aspects of the immaterial mind that go beyond the physical matter of the brain, especially for higher order cognitive and intellectual function like free will and abstract reasoning.

Does mind-brain dualism conflict with the atheistic materialist worldview?

Yes. Recognizing the mind beyond the brain conflicts with atheistic materialism, which denies anything immaterial. Mind-brain dualism recognizes the immaterial mind as distinct from the material brain. Atheistic materialists are forced to deny the existence of the immaterial mind and instead assume that all mental functions—every aspect of your experience and identity—is simply a product of the brain tissue in your skull.

How might we assess the accuracy, truth, and explanatory power of the materialist, dualist, or other mind-brain theories and worldviews?

The materialist worldview claims that the mind does not exist apart from the brain, and that the brain is the exclusive cause of activity of the mind. Therefore, discovering evidence that the mind or activity of the mind persists without the brain would disprove or seriously challenge the materialist mind-brain worldview. For example, if the mind continued to function when the brain did not, such as during documented "brain death" as recorded by medical instruments, this would disprove the theory of absolute dependence of the mind on the brain.

The materialist also claims that activity of the mind directly correlates with brain activity, and so significant damage to the brain, or removal of entire brain sections, must correspondingly damage the mind. Therefore, evidence of serious damage to the brain, or of entirely missing portions of the brain, producing no significant changes to the mind (mental activity, cognitive function) would disprove or challenge the claim of an absolute, direct relationship between brain damage and "mind damage."

Although nobody disputes that brain damage can and does damage mental activity and cognitive function, if this pattern does not always hold true it suggests deeper questions to explore about the mind-brain connection. Read the next several chapters for neuroscientific evidence that challenges the atheistic materialist worldview, including seizure experiments, brain stimulation, brain damage, conjoined twins, split-brain surgery, free will experiments, memory research, and Near-Death Experiences.

What is correlation? What is causation? What is the difference between correlation and causation?

Correlation means two events occur at the same time or sequentially. Causation means that one event causes the next event. Correlation can indicate causation, but not always. Any researcher will warn you of the danger in automatically assuming causation when you find correlation. For example, shark attacks and outdoor barbecues are correlated, but there is no causation. Instead, a third factor—summertime—links the timing of shark bites and demand for outdoor grills.

Causation is difficult to establish, and sometimes the challenge is proving the direction of causation (which event is cause versus effect). Reverse causation mixes up cause and effect, like believing that quitting smoking causes lung cancer. When lifelong smokers are diagnosed with lung cancer, they often quit smoking, introducing correlation. We know the proper causation is that smoking causes lung cancer, but looking at the data without better context could result in reverse causation, switching cause for effect. In neuroscientific research of the mind-brain connection, reverse causation and mistaking correlation for causation are serious problems that must be examined carefully.

How does the difference between correlation and causation relate to all neuroscientific research about the mind-brain connection? How does the error of reverse causation relate to mind-brain research?

Whenever neuroscientific research "proves" that the brain generates a mental activity by noticing physical changes in the brain that correlate with mental activity, be careful to accept conclusions of causation. Atheistic materialists will conclude that the brain causes any correlated mental activity, but considering reverse causation, it may instead be true that the immaterial mind causes the physical brain activity.

If your brain tissue shows electrical activity when you think (exercising your mind), correlation is established but not the direction of causation. The mind beyond the brain could prompt the brain to activate. If the brain is a physical organ assisting with conscious thought, but not controlling or creating thought, the causation starts with the immaterial mind, not the lump of brain tissue in your skull.

To the atheistic materialist, does the brain generate thoughts?

Yes. To the atheistic materialist, one function of the brain is to generate thoughts. Just as your digestive organs produce chemicals that digest food, your brain produces thoughts that create or simulate the experience of consciousness.

To the mind-brain dualist, does the brain generate thoughts?

No. To the mind-brain dualist, the mind generates thoughts through exercise of intellect and will (and for some dualists, the soul or spirit). The brain is the bodily organ that assists with certain mental activity, not the source thereof. Therefore, the brain is dependent on the mind, not the other way around.

25 Aristotle, Aquinas, and Hylomorphism

Mind-brain dualism follows the substance dualism of Rene Descartes. Minds are immaterial; brains are material. One intriguing alternative to substance and mind-brain dualism was introduced by Aristotle and refined by Saint Thomas Aquinas. Hylomorphism organizes properties (attributes), rather than substance, into the two divisions of form and matter. Therefore, hylomorphism evades many challenges of substance dualism by recognizing that all recognizable objects – including people – require both properties (attributes) of form and matter. Rather than dividing matter from the immaterial, matter needs form to assume an identity. Following hylomorphism, matter existing without form has no distinct identity or function and exists as raw potentiality until linked with the formative principle. For example, sand is potential material until it is formed into a sandcastle.

Hylomorphism corresponds well to human experience because our minds (or souls) may be explained as the formative principle of our material bodies, giving us each our personal identity as the composite of body (matter) and mind/soul (form). According to hylomorphism, our material bodies are incomplete without the living, formative principle of our minds/souls.

What are some alternative worldviews to Cartesian duality or atheistic materialism?

The property dualism of hylomorphism is a fascinating alternative to Cartesian dualism or atheistic materialism. Cartesian duality splits all reality into immaterial substances (like minds and spirits) and material substances (like the human body, physical objects, and all matter and energy in the cosmos). Mind-brain dualism—another form of substance dualism—follows this Cartesian split to distinguish the immaterial mind from the material brain. In contrast, atheistic materialism cuts out one side of the dualist equation entirely, eliminating anything immaterial: no God, no souls, no spirits, no minds. To the atheistic materialist, the mind is nothing but the brain.

Instead of dividing everything into material or immaterial substances, Aristotle and Saint Thomas Aquinas focused on the properties of the mind, body, and soul—and everything else. In this alternative worldview of property dualism, everything is analyzed according to two properties: form and matter. Form is shape or identity, and matter is substance. Matter represents raw, unformed potential that assumes its identity, shape, and function when organized by the property of form. This worldview of property dualism of matter and form is called hylomorphism.

What is hylomorphism? What is the etymology of hylomorphism?

Hylomorphism is a metaphysical worldview introduced centuries ago by the Greek polymath Aristotle (384 - 322 BC), who was a student of Plato (427 - 347 BC). The term hylomorphism comes from the Greek words for form or shape, "morphe," and matter, "hyle."

Hylomorphism proposes that matter and form are the two fundamental principles or properties of any physical object, including human beings and other living organisms. Matter is the potential to exist, like unassembled pieces. Form is the actuality that informs identity, function, and shape. Everything is a composite of matter and form, including inanimate objects, humans, and living organisms.

How does the form and matter established by hylomorphism relate to actuality and potentiality?

Hylomorphism divides everything into matter and form to distinguish potentiality from actuality. Matter represents potential, and through form, the potential matter transforms into its actualized identity, function, and shape. To Aristotle, the human soul is the formative principle that gives life and personal identity to the material body.

Under Aristotelian hylomorphism, raw material is not a finished product, so the potential must be "formed" into actuality. Therefore, the "form" of any object is just as important as the matter or raw material. Just as clay must be shaped into a sculpture, the raw material of living organisms (molecules, atoms) must be formed to give the animal or person its shape, function, and identity. If a person or a sculpture explodes into millions of pieces, the person or sculpture no longer exists. The form has been eliminated, although the matter (raw material) remains.

How does hylomorphism fundamentally differ from Cartesian dualism or substance dualism?

Cartesian dualism divides the cosmos into two substances: immaterial mind (or spirit) and material. Hylomorphism instead recognizes two properties in everything—form and matter—but these properties are inseparable parts of the composite whole. Under hylomorphism, every object is a combination of form and matter; if either form or matter is missing or destroyed, the object loses its identity, form, and function.

What philosophical issues drove Aristotle to his idea of hylomorphism?

Aristotle sought to explain how identity persists over time and despite change. To Aristotle, the property of form explains how the same individual parts of matter can be rearranged (formed) into a new identity or instead sustain over time the same form and corresponding identity.

What is the Ship of Theseus?

The Greek and Roman philosopher, historian, and biographer Plutarch (46 - 120s) is most famous for his collection of groundbreaking biographies of notable Greek and Roman leaders. Within the biography of Theseus, Plutarch was also the first to record the philosophical thought exercise known as the Ship of Theseus, which examines the nature of identity over time.

Needing repairs, the Ship of Theseus is replenished over time with newer, replacement planks of wood to preserve the ship as a historic relic. Eventually, the ship may retain only a fraction of its original material. The philosophical question posed is whether the Ship of Theseus is the same ship, retaining its identity, or a different ship, transforming into something else. As objects change material over time yet retain the same identifying characteristics, the mystery is whether or when the objects become something else entirely.

Does the philosophical issue of the Ship of Theseus apply to the human condition and sense of identity over time? Does the human body change significantly over a lifetime?

Yes. The Ship of Theseus thought exercise is highly provocative when considering personal identity over time. Physically, there is no question that the human body changes significantly over time. Every cell of your skin, red blood cells, and stomach lining is replaced completely within days, weeks, or a few months. Bone cells are replaced at a slower rate, roughly every decade. The medical consensus is that the average lifespan of every cell in your body is about seven years, although some cells (like certain neurons in the brain) do not get replaced at all. Despite the continuous biological overhaul of the material substance of your body, you probably agree that your personal identity remains the same over time. You are the same person from birth to death, even if your parts are swapped out and replaced much like the Ship of Theseus.

How does the Ship of Theseus problem relate to your worldview of the mind, body, soul, and sense of self?

Since your physical body changes constantly, your immaterial mind or soul is your only continuous identity. If you deny the existence of your immaterial mind or soul and adopt atheistic materialism, your only identity is linked with your material body. But your material body will be "swapped out" every few years with completely new cells. The ravages of aging or destruction from severe injury may further erode your limited, material identity of the body. One day you may find yourself missing entire parts of your body or burdened with an aging, decrepit, barely functional body. Only through recognition of your immaterial mind or soul can you retain a firm, consistent, indestructible identity.

What is Thomistic hylomorphism as explained by Saint Thomas Aquinas?

Saint Thomas Aquinas (1225 - 1274) built upon the hylomorphic foundations of Aristotle to explain human beings as the unified fusion of body and soul. Following Aristotelian hylomorphism of matter and form, every person is a composite of a physical body (the matter principle) and soul (the formative principle). In this way, Thomistic hylomorphism explains that the soul "forms" the matter of the body, and every individual human is identified by a unique body and unique soul. Following Christian theology, the soul is immortal and survives the death of the physical body, which is one main difference between the Thomistic versus Aristotelian conception of the human soul.

How does Thomistic hylomorphism differ from Cartesian dualism or substance dualism?

Cartesian or substance dualism divides everything—including human beings—into two substances of the immaterial (soul) and material (body). In contrast, Thomistic hylomorphism views each person as a single, composite substance consisting of the soul and body. Although the immortal soul can survive the body, both body and soul are intrinsic properties of a human being.

Notably, Thomistic hylomorphism avoids substance dualism and therefore avoids The Hard Problem of Consciousness. Under Thomistic hylomorphism, there is no mystery why the soul (or mind) and body (or brain) interact because they are two integrated principles of the same person representing form and matter.

Did Aristotle use the word "soul" the way it is understood today?

No. Today the word "soul" usually means your spirit or intangible essence, or perhaps your consciousness or mind. Aristotle used the word "soul" to describe an immaterial force acting upon organisms to govern their shape and behavior. To Aristotle, every organism is governed or formed by an invisible, immaterial soul. Under this Aristotelian theory, the soul is not within the organism, or even a part of the organism. Rather, the organism is within and subject to the soul. The soul provides a top-down influence, pulling the potentiality or matter of the organism toward its form as it grows and matures.

Did Aquinas use the word "soul" the way it is understood today?

Yes. Aquinas reconciled Aristotle's concept of souls with Christian doctrine by noting that the highest rational soul is immortal and survives death and destruction of the body. Aristotle believed that souls are merely formative principles for the raw material of the body, but Aquinas understood the rational soul as a gift from God that distinguishes mankind from the animals.

Aquinas believed that the rational soul is what connects humanity to God and is one reason why humanity was created "in the image of God" as described in Genesis, the first book of the Old Testament of the Bible. To Aquinas, human beings can connect with God and share in the divine knowledge of God by using the rational soul.

What are the three levels of souls or formative principles according to both Aristotle and Aquinas?

Both Aristotle and Aquinas proposed three types of souls as the formative principles of living organisms: vegetative (or nutritive) souls, animal (or sensitive) souls, and rational (human) souls. Plants are formed by the simplest vegetative souls that govern basic nutrition, growth, and reproduction. In addition to a vegetative soul, animals are formed by sensitive souls that provide sensation, emotion, memory, and physical movement. At the top of hierarchy, only humans have all three: vegetative, sensitive, and rational souls that form consciousness, rationality, abstract thought, intellect, and free will. For these reasons, Aristotle called humans the "rational animal," because the higher intellectual powers elevated mankind above animals.

How do the Aristotelian and Thomistic levels of souls correspond to neuroscientific research?

The lower, animal souls provide sensation, emotion, memory, and physical movement, but the higher cognitive abilities like intellect, rational thought, and free will are reserved for the formative principle of the human soul, which animals do not enjoy.

Interestingly, both brain seizures and artificial brain stimulation trigger only the functions informed by the animal soul: sensation, emotion, memory, and movement. Seizures and stimulation never induce free will or abstract thought, but only the mental activities associated with the Aristotelian and Thomistic animal soul. Of course, sensation, emotion, memory, and motion are all mental functions that humans share with animals.

Why does the inability to physically stimulate or activate the powers of the rational, human soul challenge the atheistic materialist worldview? Why should we be able to stimulate mental activities shared with animals, but not higher order activities like intellect, free will, and abstract thought?

One explanation is that the soul is immortal, immaterial, and linked to intellect, free will, and abstract thought. Atheism and materialism deny the existence of God and the immortal human soul. Yet the Thomistic and Aristotelian concept of an immortal, immaterial soul is consistent with neuroscientific evidence.

Specifically, the human soul cannot be physically, automatically activated because it is immaterial and not limited to physical brain activity. In contrast, the mental activities shared with animals are easily triggered by electrical shocks, whether from pathological seizures or when induced by researchers. Electrical stimulation of the physical brain of humans can produce strange sights or tingling sensations, sudden memory flashes, strong emotional responses, and involuntary movements—the same four symptoms of involuntary brain seizures. See chapter 27, "Seizure Experiments and Brain Stimulation." Yet neither seizures nor experimental stimulus can evoke free will or abstract thought because these mental activities are linked to the immaterial mind of the soul, not to the physical organ of the brain.

What did Werner Heisenberg notice about the connection between quantum physics and the Aristotelian ideas of potential, actuality, and hylomorphism?

German physicist Werner Heisenberg (1901 - 1976) won the 1932 Nobel Prize for the "creation of quantum mechanics." Heisenberg noticed the strange behavior of quantum mechanics confirms many aspects of ancient Aristotelian metaphysics involving potential and actuality. Aristotle's concept of hylomorphism explains that objects reach their actuality when form gives shape and function to matter which previously existed as mere potentiality. For example, clay is raw material (mere potential) before a sculptor provides the form (actuality) of the sculpture. In his 1958 book, Physics and Philosophy, Heisenberg connected the uncertainty and potentiality of the particle wave function to the actuality of position in time and space that is only discovered by observation. The connection to immaterial idealism and abstract, rational, mathematical language is undeniable:

"In the experiments about atomic events we have to do with things and facts, with phenomena that are just as real as any phenomena in daily life. But atoms and the elementary particles themselves are not as real; they form a world of potentialities or possibilities rather than one of things or facts ... The probability wave ... mean[s] tendency for something. It's a quantitative version of the old concept of [potential] from Aristotle's philosophy. It introduces something standing in the middle between the idea of an event and the actual event, a strange kind of physical reality just in the middle between possibility and reality."

26 Metaphors of the Mind and Brain: Machines and Computers vs. Televisions and Pianos

Atheistic materialism favors a mechanical, computer-based metaphor for the mind and brain. Mankind has always favored technological metaphors, often matched to the everyday technology of the time. Whatever metaphor is employed, remember that most metaphors smuggle into the discussion consequential assumptions about the mind-brain connection. Dominated by a mechanistic, materialist worldview, the computer metaphor smuggles in mechanistic, materialist assumptions about the brain controlling the mind. Yet careful examination of the correlation (not necessarily causation) between brain activity and mental activity suggests instead that the mind controls the brain, rather than the reverse. The metaphors of the piano or television for the brain may be superior to the overused metaphor of the computer, illustrating a mind-brain dualism wherein the mind "plays" the brain like a piano, or the brain receives "signals" from the mind like a television.

What is the popular metaphor to explain brain function most often used by atheistic materialists?

Atheistic materialists compare the brain to a machine, or more specifically, to a computer. Just as a computer uses energy (electricity) to compute, calculate, and run programs, the brain uses energy (from food) to compute, calculate, and produce thoughts. Your brain is like an incredibly sophisticated and complex computer.

What observed evidence supports the popular metaphor of the brain as a machine or computer?

If your brain is damaged, you risk significant deterioration of your ability to think, control your body, and perform other mental activity. If you smash a computer with a sledgehammer, the computer stops working properly. If you destroy essential circuitry, or delete critical software code, the computer will not function. In this way, damage to the brain and damage to the computer produce a similar result.

What is the inherent problem with using the computer as a metaphor for the thought process?

Computers do not and cannot think. Computers, by their very name, compute or process information. Thought is unrestricted and creative. Computers are restricted by computer programming (software, instructions) and cannot exercise creativity. Every computer follows programs written by human beings. No computer participates in thought, experience, contemplation, or understanding, nor does any handheld calculator, abacus, or other machine.

Actual thought includes the quality and experience of thinking "about" something. To think "about" something requires both subject (the thinker) and object (the topic thought about). Thought is inherently purposeful. Thought searches for and is driven by meaning. But computers cannot participate in this subjective, inherently personal quality and experience of thinking "about" any topic or idea or toward any purpose.

Why is it impossible for computers to generate independent, creative thoughts?

Computers accept input (data) that is processed through a software algorithm, resulting in a predetermined range of output (results). Computer programs are designed to process data from one form to another, but never to create new ideas, or to search for purpose, or to truly think like a human being. Many computer programs simulate or "fake" the process of thinking, or speaking, or debating, or writing, but these programs are deceptive. The only creative input for any computer program is limited to two sources: the original author of the software code that instructs the computer to run the program, and the data source for the program to process.

What are some provocative, alternative metaphors for the relationship between the brain, the mind, thoughts, and consciousness?

Piano or television. Alternative metaphors for the mind-brain interaction describe the brain as a receiver, transducer, or organ designed to process and translate information and thoughts provided by the separate mind. The immaterial mind and material brain work together, but the mind is the ultimate authority and source of higher-level thought like abstract reasoning and free will.

Consider your brain to be more like a television or piano. In the television metaphor, the brain receives, interprets, and processes external signals from the mind in the same way that a television receives signals from elsewhere.

In the piano metaphor, the mind "plays" the brain like an instrument, like how the concert pianist plays the piano. Without the piano, no music is possible, but the pianist is the source and master of the music played.

How does the television metaphor explain the mind-brain connection?

If your brain is more like a television, your brain receives signals from elsewhere and processes or transduces (translates into another medium) those signals. The television receives programming signals using an antenna, internet connection, or a cable programming subscription.

Next, the television converts (transduces) that signal into a visual representation (pixels on a glass screen) and an audio output (sound coming through speakers). The television is a receiver and transducer (translator).

In the television-brain metaphor, the brain receives signals from elsewhere, such as from your mind, your consciousness, or even from other conscious entities. Then, your brain transduces (translates into another medium) those signals. The key to the television-brain metaphor is that just like the television is not the source of images on the screen, the brain is not the source of all mental activity. Instead, both the television and brain receive and process external information and signals.

If the television is damaged, the output (audio and visual) could be distorted or stopped entirely. If you remove the necessary wiring or circuitry, the television cannot receive or transduce the signal. If you remove the television speakers, the television cannot produce sound. If you crack or remove the glass screen, the television cannot produce moving pictures to view. If the television receives no signal at all, there is no sound or video for the television to produce for your viewing pleasure. Similarly, brain damage may affect mental activity. But in this comparison, neither the brain nor the television is the source of information, but rather, both are the receiver and translator.

What is the piano metaphor of the brain?

The piano is a musical instrument, played by the musician to produce musical notes and melodies and songs. The piano must be tuned correctly for the sound of the notes to be correct, harmonious, and beautiful. If the piano is poorly tuned, the music the piano produces will be degraded and distorted. If you smash or break the piano, the musician may not be able to play the piano at all, and no notes or melodies or songs are possible. If the musician cannot touch the piano, or if the piano is separated from the musician, the pianist cannot produce any music via the piano. Similarly, the quality or health of the brain may affect mental activity even though the brain is not the source of mental activity.

In the piano-brain metaphor, the pianist represents your immaterial mind. Your mind is the musician to "play" your brain like an instrument. A brilliant pianist will struggle to squeeze beautiful music out of an inferior piano, but will dazzle the crowds with a prestigious, properly tuned Steinway & Sons, Yamaha, or Fazioli-brand grand piano. In the same way, a superior mind is related to but not entirely dependent on the physical instrument of the brain. Nobody doubts the interplay of musician and instrument, nor should we doubt the interplay between the mind and the brain. But the piano-brain metaphor completely reverses the materialist worldview: the mind "plays" and controls the brain like an instrument, rather than the brain being in control.

How do the metaphors of the television and piano correspond to observed evidence about the mind-brain relationship?

Seizure studies and experiments, split-brain patients, Near-Death Experiences, conjoined twins, and patients with severe brain damage all provide compelling evidence that mental activity is not entirely dependent on the physical brain. The next several chapters will explain the following challenges to the atheistic materialist worldview of mind-brain interaction:

1. Seizures never engage higher level thought like abstract reasoning or free will. Brain seizures are limited to automated reactions of movement, memory, emotion, and perception—the same reactions that can be provoked by physically prodding the brain.
2. Physical stimulation of the brain never activates higher level thought, but only lower functions like body movement, memory recall, emotional urges, and distorted perception. Free will and abstract thought cannot be activated by physical prodding.
3. People born with entire regions of the brain missing, and those who lose sections of the brain from accident or necessary surgery, persevere to perform the same mental activities as those with complete brains. The plasticity or flexibility of the brain regions to perform different tasks suggests an integrative system of functioning and poses a particular challenge to reductive, materialist worldviews.
4. Conjoined, identical twins joined at the head and sharing one brain nonetheless have two distinct minds. Two brains can be combined, but not two minds into one mind.
5. Split-brain patients show remarkably little change in mental activity after this invasive surgery. The brain can be split in half, but the immaterial mind is indivisible.
6. Urges and impulses—and perhaps temptations—show up on brain scans, but the exercise of free will is not detected by material instruments. The exercise of free will may therefore transcend materialist explanations and be sourced from the immaterial mind, not the brain.
7. Despite decades of painstaking research, no physical location of memory storage in the brain has ever been found. Even when significant chunks of the brain are removed, memories often persist, no matter the location of physical loss.

8. Near-Death Experiences have accumulated thousands of cases, with hundreds involving verified information perceived by someone who is clinically dead, brain dead, blinded by surgical tape over the eyes, or heavily anesthetized. There is no way for the eyes, ears, or brains to have perceived what these people recounted upon returning to normal consciousness. For example, many Near-Death Experiences involve seeing your own body and the operating surgeons from an elevated view outside the body and then being able to recall and share the precise sights and sounds to the astonished confirmation of medical professionals. None of this should be possible if the mind is nothing but a product of the brain, nor if consciousness is limited to brain activity.
9. The placebo effect demonstrates that the immaterial mind has powerful control over the health and healing of the body, despite no mechanical or material explanation.

27 Seizure Experiments and Brain Stimulation

Seizure and brain stimulation research provides compelling evidence that the mind exists beyond the brain, defeating a key assumption of atheistic materialism. While the brain can experience seizures or artificial stimulation leading to unexpected movements, perceptions, emotions, or memories, none of the higher mental activities like reasoning, abstract thought, or free will can be triggered by physical impulses. Nor has anyone ever been convinced that these involuntary reactions to artificial brain stimulation came from free will rather than from careful prodding for scientific research. People can distinguish between their internal free will versus involuntary, external brain stimulation. The obvious inference is that reasoning, abstract thought, and free will are qualities and functions of the immaterial mind (or soul) rather than the brain.

What is an epileptic seizure?

The chronic condition of epilepsy results in repeated seizures spurred by abnormal electrical charges in various parts of the brain. Although epileptic seizures cause a variety of symptoms, the most familiar result of a seizure is rapid, involuntary body movements.

What did neuroscientist Wilder Penfield discover about epileptic seizures?

Neuroscientist and neurosurgeon Wilder Penfield (1891 - 1976) systematically studied the effects of epileptic brain seizures, and he noticed there were four and only four reactions: involuntary movement and spasms of the body, false perceptions like tingling sensations or light flashes, powerful emotions like dread or joy or laughter, and invasive memory recollections. Penfield drew the connection between the seizure reactions and four essential brain or mind functions: movement, perception, emotion, and memory. Penfield thought it strange that seizures never induced higher cognitive activities like abstract reasoning or free will or rational thought, but only those four mental functions.

How did Wilder Penfield map the four brain functions (revealed by seizures) to different areas of the physical brain?

Penfield knew that seizures were caused by sudden electrical sparks or surges in the brain. Therefore, Penfield thought that he could also stimulate the brain, manually, to induce the same movement, perception, emotion, and memory reactions. Indeed, Penfield's brilliant experiments on "open brains" (involving more than 1,000 healthy test subjects with their skulls cut open) revealed that if you physically stimulate different parts of the human brain, you can create the same four reactions caused by seizures: movement, perception, emotion, and memory. Notably, Penfield failed to find any brain locations tied to abstract thought or reason.

What is one inference from Wilder Penfield's failure to locate any region of the brain that can be stimulated to produce abstract thought or reason?

The fact that Wilder Penfield could poke and prod the brain to produce movement, perception, emotion, and memory—but not abstract thought or reason—suggests that abstract thought or reason is not reduced to mere brain function. Higher cognitive activity, abstract thought, reasoning, and free will could instead be powers of the immaterial mind that exist beyond physical brain function. This inference of separate function from the brain also corresponds to the fact that involuntary seizures stimulate only movement, perception, emotion, and memory, but not abstract thought, reasoning, and free will. Through both internal brain seizures and artificial, physical stimulation of brain tissue, these four and only these four mental activities were triggered.

How did Wilder Penfield further distinguish the effect of free will from the involuntary stimulation of movement, perception, emotion, and memory? Were open-brain test subjects able to distinguish between their free, voluntary movements versus limb movements artificially stimulated by Penfield?

Penfield was curious about free will because he was not convinced that all aspects of the mind were completely and physically dependent on the brain. To test the effect of free will, Penfield asked his test subjects to occasionally move their limbs using their own free will, rather than as a passive reaction to Penfield's artificial, physical stimulation of their exposed brain tissue. Brains have no sensory perception of touch, and the patients' eyes were blinded from Penfield's intervention, so no sense of sight or touch could reveal whether Penfield intervened. In this way, Penfield investigated whether the test subjects could mentally distinguish the movement of their body based on free will rather than an involuntary movement prompted by Penfield's stimulation.

Can your mind distinguish between your own voluntary movements versus artificial stimulation of your brain that causes involuntary movements?

Yes. In Penfield's experiments on more than 1,000 open-brain test subjects, the results were overwhelmingly consistent: every person could distinguish between moving limbs by free will versus the feeling that the limbs moved because of outside, artificial, physical stimulus from Penfield. Somehow, the test subjects knew whether their bodily movements were freely chosen as opposed to physically forced by external meddling, which seems to confirm the existence of free will, independent from physical brain stimulation.

28 Brain Damage and Missing Brain Regions

If the mind is nothing but the brain and atheistic materialism explains everything, there should always be a direct, one-to-one correlation between brain damage and mental function. Yet the evidence shows otherwise. Millions of people have suffered significant brain damage or were born missing entire regions of the brain, yet their mental function is relatively normal. Nobody denies that brain damage can reduce mental function, but the surprising fact is how adaptable and flexible the mind can be despite severe physical brain damage. Evidence of the mind persisting and thriving despite physical brain damage suggests that not all mental activity can be reduced to physical brain matter.

What evidence proves that the direct, absolute relationship between brain damage and "mind damage" is not as well-established as claimed by mind-brain materialists?

Neuroscientists have performed a wide variety of drastic surgeries on the human brain. You would not be surprised to hear that removal or mutilation of major sections of the brain can and does affect both brain function and activity of the mind. Yet you would be surprised to learn of many examples of drastic brain surgery that show the opposite result: no changes to mental activity despite significant removal or destruction of brain tissue, including entire regions or brain sections.

There are also countless people born with entire regions of the brain missing, including those considered essential to major mental functions, yet many of these people have completely normal lives with little to no effect on mental activity. Of course, a damaged or missing brain often results in severe limitations to mental activity—nobody denies the potential harm of brain damage. Yet the many cases wherein mental activity persists despite severe brain damage or "incomplete" brains with entire regions missing undermines the materialist belief of a direct, one-to-one correlation of physical brain tissue and mental activity.

What is one explanation for the persistence of relatively normal mental activity of those who have missing brain sections, severely damaged brains, and surgically altered brains?

Mind-brain dualism or the worldview that the immaterial mind exists beyond the material brain explains why mental function may persist despite an incomplete or damaged brain. If the mind controls the brain as an organ during the process of thinking, then not all brain damage or disfunction will preclude mental activity. The mind might be able to adapt to using the organ of the brain differently depending on what is available to use. If instead the brain were simply a machine that produces the mind, we would expect a universal, one-to-one ratio of brain damage (or removal) to mental function decline or failure. Yet the evidence supports a much more flexible and adaptive relationship between the mind and brain, pointing to mind-brain dualism, and conflicting with the worldview of reductive, mechanistic materialism. Furthermore, the plasticity and interchangeability of physical brain regions to perform different mental functions also gives compelling evidence against a strict materialist interpretation of the mind-brain connection.

29 Conjoined Twins Sharing One Brain

If the mind is nothing but the brain, then one brain can produce only one mind. Yet conjoined twins who share one brain mass have two completely different minds with distinct identities, personalities, consciousnesses, and memories. If instead the mind controls the brain as a bodily organ involved with mental processes, two minds can share one brain just as two bodies (of conjoined twins) can share other vital organs like lungs or kidneys.

Do conjoined twins who share a brain have two different minds and two different consciousnesses?

Yes. Conjoined twins who share a single, connected mass of brain tissue have two different minds and two different consciousnesses. Under the atheistic materialist worldview of the mind as nothing but the brain, you would expect a single, shared brain to produce a single, unified mind and consciousness. Yet conjoined twins who share a single, connected brain mass are two distinct people, with different personalities, desires, free wills, and consciousnesses.

What are conjoined twins?

Conjoined twins (once called Siamese twins) are an extremely rare type of identical twins who are born physically connected, and sometimes sharing organs or body parts.

What do conjoined twins who share brain matter suggest about the mind-brain connection and the nature of consciousness?

Conjoined twins who share brain matter pose serious challenges to the reductive, materialist, mechanistic worldview of the mind-brain connection.

Under the materialist worldview, one brain should produce one mind and one consciousness. Yet twins conjoined at the head, sharing a single, connected brain mass, exhibit two minds and two consciousnesses, suggesting that immaterial minds exist beyond the brain.

Under the materialist worldview, how many persons, minds, and consciousnesses should we expect when limited to a single brain?

The materialist worldview expects a single person with one mind and one consciousness derived from one brain. Instead, two minds of conjoined twins operate distinctly yet share the use of a single brain mass, suggesting that the immaterial mind is more than physical brain function.

How do conjoined twins who share brain matter indicate that minds use brains as tools? Do these conjoined twins totally reverse the materialist explanation of causation between the mind and the brain?

If the mind is limited to brain function, and the brain produces consciousness, we should expect a one-to-one ratio of minds to brains. Yet the two minds of conjoined twins share a single brain mass, using that brain more as a tool. Somehow, two minds operate distinctly while sharing access to and use of the same, single brain mass. For example, you can explore the sensory, perception, identity, and consciousness anomalies evidenced by conjoined-at-the-head twins Krista and Tatiana Hogan (born 2006).

The mind using the brain as a tool or organ to assist in mental activities directly contradicts and reverses the reductive, materialist notion of the brain being the primary source and "control center" for the mind. Twins conjoined at a single, shared brain provide powerful evidence that immaterial mind, consciousness, personality, and free will operate outside the limits of the physical brain.

30 Split-Brain Surgery Does Not Split the Mind

If the mind is nothing more than the brain, cutting the brain in half should severely impair mental function. Yet neuroscientific evidence proves that patients with split brains have little to no damage to their seemingly indivisible minds. Mechanistic, materialist metaphors of the brain completely break down when confronted by the fact that cutting the brain in two barely affects the immaterial mind. Unlike the indivisible, immaterial mind, machine metaphors for the brain fail because mechanical contraptions cannot function after being sawed in half.

Can you cut your brain in half and retain the unity of your mind?

Yes. Split-brain surgery proves that you can cut your brain in half but retain the unity of your immaterial mind. While the physical brain can be divided, your mental function remains indivisible. Indivisible unity is an expected property of the immaterial worldview of the mind and contradicts the materialist worldview.

What is split-brain surgery?

One method of brain surgery to alleviate seizures is to cut the brain in half, totally separating the left and right hemispheres (sides) of the brain. The corpus callosum is a bundle of nerve fibers that connects the two brain hemispheres. Seizures often start in one section of one side of the brain and then spread throughout the entire brain. By completely slicing through the corpus callosum, thereby splitting the brain in half, neurosurgeons can stop or reduce the severity of the seizures. But the shocking result is that in most cases of split-brain surgery, the patient has no fundamental change in mental activity (other than suffering from fewer or milder seizures).

How did neuroscientist Roger Sperry investigate split-brain patients to reveal anomalies about the mind-brain connection?

Neuroscientist Roger Sperry (1913 - 1994) realized that split-brain patients can be studied to understand the connection between the left and right hemispheres of the brain. Sperry discovered many insights about the interaction and function of the two brain halves.

For this discussion of mind-brain interaction, the most interesting result of Sperry's exhaustive investigations is that patients' mental abilities (minds) were completely normal although their brains were literally split in half, with only some minor reductions to perception or motor ability. Therefore, one inference is that despite the physical brain being divided into two pieces, the immaterial mind is indivisible and can persist despite severe physical trauma to the brain. Perhaps the physical brain can be divided, but the mind cannot. Material can be divided, but immaterial minds are inherently unitary.

How did neuroscientist Justine Sergeant discover evidence of the indivisible, unitary mind despite a physically divided brain?

Neuroscientist Justine Sergent (1950 - 1994) experimented with split-brain patients by presenting different pictures to the right brain and to the left brain and then asking the patient to distinguish the two. For example, Sergent would present a picture of an arrow facing left to the one brain hemisphere, and an arrow facing up to the other side of the brain, and then ask the patient if the arrows pointed in the same direction.

Although the split-brain patients had brains that were completely disconnected by being literally cut into halves, somehow the patients could distinguish and compare the two pictures. Yet no part of the brain saw both images. Instead, some mysterious process of integration allowed for the comparison of the two images. Sergent's work suggests that despite the perceptive function being divided in two (by cutting the brain in half), the abstract reasoning or rational function of the mind remained indivisible and was able to integrate the two disconnected signals from two disconnected brain-halves despite no part of the brain receiving both signals.

How did neuroscientist Yair Pinto confirm Justine Sergent's experiments suggesting that abstract thoughts and reasoning can remain united and undivided, despite the literal division of the brain and physical perception?

Neuroscientist Yair Pinto of the University of Netherlands expanded upon the work of Justine Sergent. Pinto showed two different images to the two disconnected hemispheres of split-brain patients. Only through viewing both images and using abstract reasoning could anyone ascertain the "story" or meaning of the pair of images. For example, Pinto could show a picture of a baseball to one half of the brain, and a picture of a broken window to the other half. Somehow, Pinto's test subjects could integrate both pictures in their minds despite their split brains having no way to physically share the pictures or look at both pictures at the same time.

Like Justine Sergent's results, despite no part of the physical brain seeing both images, and despite a literally disconnected brain, these split-brain patients could solve the puzzle by mental integration. The possible inference from both Pinto's and Sergent's work is that abstract reasoning powers of the immaterial mind remain unified and indivisible despite a literally disconnected brain that is sawed in half, with no pathway for physical communication.

What is one non-materialist property of the disembodied, intangible mind?

Minds are supposed to be indivisible, unlike material objects. Following a non-materialist worldview that allows for immaterial minds, the mind is believed to be inherently unitary, with no distinct parts. Although the mind can be integrated into a body, the immaterial mind is always an indivisible unity. Despite fracturing, halving, separating, or dividing the body (or brain), the mind cannot be divided, but remains intact and whole. This non-materialist worldview of the immaterial, indivisible mind also aligns with neuroscientific evidence.

31 Free Will, Free Won't, and Benjamin Libet

Pioneering neuroscientific research by Benjamin Libet shows that brain scan activity correlates with free will combined with voluntary physical movement. When the test patients acknowledged the desire to touch a button, Libet recorded brain scans that showed a spike in brain wave activity immediately prior. The controversial question is whether the brain scan activity represents the exercise of free will (as atheistic materialists believe), or rather, the initial impulse to which free will either indulges or denies (according to the mind-brain dualist worldview of the immaterial mind beyond the brain).

Intriguingly, no further brain activity registered when the test patient vetoed or denied the desire to touch the button at the last second. In other words, the brain scan activity could be interpreted to register the physical impulse in the brain, but not the mental decision to act upon or ignore the impulse, because the immaterial mind is beyond the physical brain. This mind-brain dualist interpretation aligns with the immaterial mind and free will exercising restraint and authority over a constant barrage of physical desires, impulses, urges, and temptations. Libet himself believed that his brain scans at least proved "free won't" if not free will, but atheistic materialists interpret the same data to conclude that the physical brain generates decisions and free will is illusory.

Can we detect free will or intentions using brain scans? How should we interpret electrical activity of the brain that correlates with activity of the mind?

Brain scans consistently show brain activity around the same time as reported decisions, intentions, or impulses. The persistent mystery is how to interpret the electrical activity of the brain.

Correlation is not causation, and an independent mind may stimulate brain activity, rather the brain controlling the mind. In other words, the interpretation of brain scans is just as important as the detection, and your metaphysical worldview strongly influences any conclusions.

How do atheistic materialists interpret brain scan activity correlated with mental activity?

Based on a mechanistic, reductive worldview, most neuroscientists see electrical activity in the brain and immediately assume it explains if not entirely encompasses mental activity. The atheistic materialist worldview usually interprets brain scan activity as the source, origin, and entirety of mental function, assuming that the brain directs the mind. However, correlation does not prove causation, and the direction of causation is notoriously difficult to establish in neuroscience.

How can the same brain scan activity invite a completely different interpretation?

One alternative explanation reverses the order of causation, holding that the independent mind activates the brain, which is then detected by brain scans. Materialist neuroscientists may be "putting the cart before the horse" by assuming that the brain controls the mind, rather than the reverse. The difficulty lies in the fact that immaterial causes are ignored or downplayed when using physical instruments that are designed to exclusively detect physical phenomena. If you constrain your experiment to detect material events, and restrain your analysis to mechanistic materialism, the conclusions are limited from the beginning.

How did neuroscientist Benjamin Libet use physical brain scans to discover insights about the interaction of free will, intention, and deliberate actions?

Benjamin Libet (1916 - 2007) was interested in how free will, intention, and deliberate actions can be measured by brain activity. Libet designed an ingenious experiment whereby subjects were told to signal right after they manifested the desire to press a button on a table, and therefore right before they physically pressed the button. This experiment used electrodes taped to the skull to measure brain activity of the patient's intentions and resulting action.

The time between the perceived intent to press the button, and when the patient physically pressed the button, was measured on average to be 200 milliseconds, or two tenths of a second. The surprising result was that the patient's brain activity always spiked on average 300 milliseconds before the perceived intent to act. In other words, the brain delivered a recordable, unmistakable signal right before the patient perceived and manifested the intent to press the button.

How did Benjamin Libet try to isolate what he called "free won't" as opposed to free will?

Libet further refined his experiment to explore free will by asking his test subjects to veto or suddenly reject the decision to press the button. The test subjects would decide to press the button, but then suddenly deny the impulse as an exercise of free will, or rather, "free won't."

Strangely, Libet detected no additional brain activity in subjects who exercised "free won't" by rejecting the initial decision to press the button. Subjects who carried out the initial decision to press the button, and subjects who rejected the initial decision and did not press the button, exhibited the same brain activity (a surge 300 milliseconds before manifesting the original intention).

To summarize, Libet detected physical brain activity just moments before the exercise of free will. Whether or not the initial impulse (shown by recorded brain activity) to press the button was carried out or ultimately denied, there was no further recorded brain activity. These results are fascinating and provocative no matter how they are interpreted.

What are the various interpretations of neuroscientist Benjamin Libet's discovery of brain activity before conscious intention? What is the materialist interpretation of brain activity preceding conscious intention? What is the mind-brain dualist interpretation of free will responding to brain activity?

Materialists believe that Benjamin Libet's discovery of brain activity before conscious intention means there is no free will, and humans simply respond to brain activity. Under the materialist worldview where the mind (and free will) is only the product of brain activity, the Libet experiments are touted as confirmation. The test subjects reacted to recorded, physical brain activity moments before a decision of free will, whether the decision was carried out or ultimately vetoed (denied). However, the more intriguing and robust explanation is that Benjamin Libet's experiments proved that free will exists.

The mind-brain dualist worldview corresponds precisely to Libet's findings, especially when also considered within the traditional understanding of temptation. First, mind-brain dualism is confirmed because the initial brain impulse to physically move the hand to press the button was detected, but no brain activity indicated the exercise of free will (because the mind is immaterial). In both the exercise of free will to engage in physical action and to deny the impulse to act, there was no physical brain activity detected at all. The only recorded brain activity came from the initial impulse or urge or temptation to movement—preceding free will—not from the later exercise of free will, deciding to carry out or deny the movement. If free will is superior to brain function, and free will can instead control and override impulses from the brain, Libet's findings confirm this worldview in profound ways.

How do the neuroscientific findings of Benjamin Libet correspond to traditional religious and spiritual worldviews of temptation, desire, urges, and sin?

Following the traditional understanding of temptation, our minds (and perhaps brains) are continually bombarded with impulses to act. Because we are made of flesh and blood like animals, we are subject to basic, animal desires. Some of these desires are sinful and selfish, like lust for sexual desire, gluttony for tasty food, and greed for material possessions. Yet because we are not just flesh and blood, but created in the image of God, we are endowed with free will, rational thought, and a spirit or soul.

Through the exercise and refinement of our spirit or soul, we can use free will and rational thought to accept or deny the baser instincts, impulses, and temptations. For example, you may feel a physical temptation to overindulge in pumpkin pie, but your rational thought intervenes with your long-term desire to be healthy, and you exercise free will to resist the urge. Your free will and rational thought supersede the physical temptation or impulse of your physical body. In this way, Benjamin Libet's experiments may confirm the traditional religious worldview with astonishing accuracy. The brain scan detected the initial urge or impulse in the brain, but no instrument can detect the immaterial mind and free will accepting or denying the physical temptation, which explains why no further brain scans were detected beyond the initial physical urge.

32 Can You Find Memories Stored Physically in the Brain?

Ask an atheistic materialist where human memories are stored and the confident answer will be "in the brain" despite the lack of evidence for this assumption. Despite decades of neuroscientific research, including invasive and probing surgical alteration and artificial stimulation, there is no confirmation of any specific location in the brain where memories are stored. The mechanistic belief that memories are physically inside the brain is a baked-in assumption along with the worldview of atheistic materialism that reduces all mental activity to the brain. Faith in the materialist worldview, not evidence from neuroscience, fuels the promissory hope that one day the physical location of memory storage in the brain will be discovered.

Are memories stored physically in the brain?

Not according to any definitive research findings. Evidence from experimentation and research suggests instead that memories are not stored physically in the brain, or not in specific locations. Yet the mainstream consensus is that memories are stored physically in the brain, mostly due to materialist assumptions rather than concrete evidence.

What is the difference between memory storage and memory access? How might the brain be used for memory access rather than memory storage?

Memory storage refers to the physical location or media that holds a memory. In contrast, memory access is the activity of recalling, receiving, or retrieving the memory. For example, when you access electronic files using the internet, your computer performs the access function, but not the storage function. Instead, through your internet connection, your computer retrieves files that are stored on servers in a different location. Those files are transferred (downloaded or temporarily accessed) to your computer, but the files did not originate from your computer. Some propose that the brain and memories work in the same way. Although the brain is the organ used for accessing memory, the memory is stored elsewhere, even outside the brain.

If you ask most neuroscientists, will they explain that memories are stored physically in the brain?

Yes. Despite the absence of conclusive evidence, most neuroscientists will tell you that memories are stored physically in the brain. However, this belief or assertion is an assumption based on the materialist worldview, not an observation based on experimentation and analysis.

Why do most neuroscientists simply assume that memories are stored physically in the brain?

The assumption that memories are stored physically in the brain follows the prevailing scientific worldview, which is atheistic materialism. If the entire universe is nothing but material, and our brains and bodies consist of matter, then our memories must be stored somewhere, physically, within our body. The brain—long recognized as an organ involved with thought processes—seems to be the best candidate for memory storage. Therefore, the materialist assumption is that memories must be stored physically in brain tissue.

What does neuroscientific research prove about memory access and recall—but not necessarily storage—within the brain?

The fact that both seizures and artificial stimulus of brain tissue can force you to involuntarily access and recall memories establishes that the physical brain plays a role in memory access, but not necessarily memory storage. (See chapter 27, "Seizure Experiments and Brain Stimulation.") Just as a television or computer can access signals or files not stored internally within the device, it is possible that the brain is an organ that accesses memories stored externally.

What is the term for the location of where memories are supposedly stored in the brain?

The term "engram" describes the physical location within the brain where memories are stored. Yet the search for this elusive engram continues with no definitive answers. Scientists have not confirmed the engram's location or existence, and myriad experiments suggest the entire concept of an engram is flawed. Perhaps memories are not stored physically in the brain.

What are some different types of memory? What is recognition, recall, habitual (process) memory, and episodic memory?

The simplest type of memory is mere recognition. When you see someone familiar, even if you do not recall his name, you instantly recognize that person. If you watch the same movie twice, you may recognize that you have already watched the film, even if you cannot remember every detail of the storyline.

Beyond mere recognition, memory recall requires a certain amount of specificity. The difference between merely recognizing someone and recalling the first name makes all the difference in a social setting. Similarly, you may recognize a building but not recall further details about the inside of the building. You simply know that you have seen that building before. Therefore, memory recall is more complex and detailed than mere recognition.

Another type of memory is habitual memory or process memory. You learn a new skill by following directions or by mimicking others. Through repeated practice, you commit the routine to memory. Eventually the process becomes habitual, unconscious, or automatic. You do not have to recall or think about what to do next; you have completely memorized the process or unconsciously adopted the habit. Some trainers call this experience "muscle memory" to indicate that your muscles and limbs go through the motions automatically, without conscious direction.

Episodic memory is when you commit an entire sequence, story, or incident to memory. Episodic memory is rarely complete, but it is always multifaceted. You remember a series of steps or experiences or a coherent storyline.

How has humanity used metaphors to speculate how or where memories are stored?

Over the course of human history, various philosophers, theologians, scientists, and technologists have proposed a wide variety of metaphors to explain memory. One striking pattern is that new metaphors often correspond to the technology of the time. Therefore, these technological metaphors smuggle in the assumption that human memory operates in a physical, mechanical process, like a contraption.

How did Plato imagine the human memory process?

In Ancient Greece, one familiar technology was the seal ring, containing an engraved design that corresponds to the identity of the owner. The hard, engraved image on the face of the ring was pressed against soft, melted wax. When the wax cooled, the image of the seal was impressed upon the hardened wax. Using this technique, you could mark or sign letters or other objects with an image from your seal ring.

The Ancient Greek philosopher Plato (427 - 347 BC) imagined that memories are impressed upon the mind like a seal ring impressed upon wax:

"Let us, then, say that this is the gift of Memory, the mother of the Muses, and that whenever we wish to remember anything we see or hear or think of in our own minds, we hold this wax under the perceptions and thoughts and imprint them upon it, just as we make impressions from seal rings; and whatever is imprinted we remember and know as long as its image lasts, but whatever is rubbed out or cannot be imprinted we forget and do not know."

How did the early 20th-century psychologist, Thomas Hatherley Pear, imagine the human memory process?

In 1922, the British psychologist Thomas Hatherley Pear (1886 - 1972) compared human memory to the technology of the gramophone or record player. At that time, the gramophone was state-of-the-art technology for playing music that was recorded on circular disks called records.

To record music, the physical vibrations of musical sounds are translated into cuts and grooves on the disk. The needle or stylus of the gramophone traces these disk grooves, which vibrates the needle and reproduces signals for frequency, pitch, and tone. Based on the vibratory feedback of the needle as it traces the path of the record's grooves, the gramophone amplifies these sounds, resulting in audible music for the listener.

T.H. Pear imagined that human memory is similarly etched into the mind, the same way the vibratory patterns of musical sounds are cut into the grooves of a record. Pear compared the quality of the memory recall to the mechanical quality of the record player. If the needle or stylus were not sharp enough, the music would replay with less fidelity to the original recording. If the record's patterns were not precisely cut, the quality of the musical output would suffer.

Similarly, Pear believed that our memory recordation and our memory recall or playback could suffer based on defects in the mind or its processes. Just as a blunt needle would stifle the music played by a gramophone, a dull or damaged mind would fail to properly recall memories. Just as a damaged disk would fail to produce music when placed in the gramophone, human memories could degrade over time or due to trauma.

How did the famous psychologist Sigmund Freud imagine memory?

Psychologist Sigmund Freud (1856 - 1939) likened the human mind to an architectural house. The human body is the exterior of the house, and the mind is the interior. The various rooms, hallways, and storage spaces of the house represent different aspects of the mind.

To Freud, memories could be readily available—stored in accessible locations in the entranceway, for example—or they could be hidden. Repressed memories might be tucked away in the attic or even screened or covered by other memories, to avoid the repeated trauma of recalling painful memories. Freud's memory metaphor of the house once again draws upon a physical, material, technological comparison.

How did the invention of the computer affect common metaphors for human memory?

Whether you consider the inventor of the first computer to be Charles Babbage, Alan Turing, Konrad Zuse, or anyone else, the emergence of computer technology immediately invited metaphorical comparison to human memory, and to other thought processes of the mind. The computer records and stores information. The computer recalls data and can present it to the user. The computer processes information and performs calculations. All these functions have obvious similarities to how we perceive the workings of the mind.

How does the long-term and short-term memory storage in computers relate to human memory?

Computers have both long-term and short-term memory. Long-term memory is stored in the hard drive. Short-term memory is found in Random Access Memory (RAM) chips. The comparison to human memory is obvious. We distinguish between long-term memory, available for recall many years later, versus short-term memory, often forgotten within minutes or hours. You remember where you parked your car for a few hours, but you forget the location when you drive away. But you remember the name of your friends and family for years or for your entire life. In this metaphor, your parking spot is stored like RAM in the computer, and your friend's name is stored for the long-term in the mind, like the hard drive storage of a computer.

What experiments did Karl Lashley conduct to try to isolate the location of memories in the brain?

American psychologist Karl Lashley (1890 - 1958) performed a series of experiments designed to locate the engram, or physical location of stored memories in the brains of rats. Like many other experiments on rodents, the hope was that knowledge about rat brains would inform our understanding of human brains as well.

In the early 20th century, Lashley trained rats to navigate mazes. Once the rats committed the route to memory, demonstrated through flawless navigation of the maze, Lashley cut out portions of the rats' brains. The idea was that if Lashley cut out the part of the brain that (supposedly) stored the memory, the rats would fail to remember the maze route. If Lashley cut out some unrelated parts of the brain, the rats would retain the memories and successfully navigate the maze.

In Lashley's famous experiment, did the rats remember the maze route even after having entire portions of their brains removed?

Yes. Surprisingly, the rats remembered the maze route, completing the circuit flawlessly, even after having entire portions of their brain removed. The goal of isolating and identifying the elusive engram failed. Lashley's experiments failed to find a definite location within the rat-brain that stores memories of the maze.

Shockingly to Lashley and others, no matter which parts of the brain were removed, the rats retained the memory of the maze, unless the brain was so damaged or destroyed that nothing functioned at all. Any observed memory loss was correlated to the amount of the brain removed, but not to the location of the removed tissue.

What do the Karl Lashley rat-brain experiments suggest about how the brain supports memory?

Because the memory loss was not correlated to any specific location and was instead related to the degree of damage to the entire brain, the Lashley experiments suggest that memories are not isolated to a specific region of the brain. Instead, Lashley's experiments opened the possibility of "mass action" of the brain—the hypothesis that the entire brain, or several regions of the brain, are responsible for memory storage and memory recall. These findings suggest a more comprehensive or integrative process of memory retention and recall, differing sharply from materialist assumptions about physical, discrete storage locations.

What else did the Karl Lashley rat-brain experiments suggest about how the brain supports memory?

Lashley's experiments also revealed another surprising feature of the brain-memory interaction. Despite cutting away various and diverse sections of the brain, the rats retained the memory of completing the maze routes. Undamaged parts of the rats' brains seemed to compensate for or take over the functions of the damaged regions. This phenomenon was called "equipotentiality" to suggest that different parts of the brain have equal potential to perform the same functions, especially when presented with the need to compensate for lost or damaged brain tissue. This "equipotentiality" indicates a strong plasticity or adaptability of the brain.

Has the "equipotentiality" of the brain, especially for memories, been demonstrated beyond the experiments of Karl Lashley?

Yes. Many other experiments produced similar results wherein cutting out a variety of different parts of the brain fails to erase memories. The "equipotentiality" of memories has been demonstrated in the memory of baby chicks, octopuses, and in a wide variety of other mammals, reptiles, insects, birds, and even human beings.

What is another word for the "equipotentiality" of the brain?

Another word for "equipotentiality" is plasticity of the brain, suggesting a flexibility of function and location throughout the brain.

Has "equipotentiality" and the plasticity of the brain been demonstrated in human brains for a wide variety of mental activities, not just memory recall?

Yes. Many human subjects have demonstrated that despite severe brain damage in certain regions of the brain, other locations of the brain can compensate for or take over the functions previously performed by the destroyed or missing portions. For example, when patients with a brain tumor have entire chunks of their brain removed that are associated with balance and coordination, these patients often recover completely. Over the course of weeks or months, the plasticity or adaptability of the brain allows the remaining portions of the brain to take over the function of balance and coordination. This plasticity and adaptability of the human brain is well-known and well-observed for a variety of mental functions.

What are alternative hypotheses or explanations for the results of the Karl Lashley rat-brain experiments and the phenomenon of brain plasticity?

Some hypothesize that rather than being stored in a specific location, memories are stored everywhere throughout the entire brain. The best metaphor for this theory of memory storage in the brain is the hologram, or holographic theory. If you cut off a chip or piece of a hologram, the entire image is available within that segment. In the same way, the holographic theory of the mind and brain supposes that each memory is available in every part of the brain, thereby explaining why memories persist despite the removal of entire regions of the brain.

Another alternative theory is that memory is not stored physically in the brain. Rather, the brain is an organ involved in accessing or receiving information like memories, which are stored elsewhere. Whether the outside storage of memories is in the intangible mind, or found within some other dimension or medium, raises entirely different questions and challenges to the materialist worldview.

How did Aristotle think of the process of human memory?

The ancient Greek philosopher Aristotle (384 - 322 BC) believed that memory was a form of perception. Through memory, we peer into and view the past. Rather than directing our attention to some stored memory, Aristotle considered memory to be direct perception or engagement with past events or details.

How do modern memory retention practices align with Aristotle's view of memory as perceiving or peering into the past?

You may have heard that memory experts recommend that you imagine an entire house filled with the knowledge you want to memorize. You place the first bits of knowledge in the entryway, and you imagine a distinctive item in that room of the imaginary house. The next thing you should remember is stored in the kitchen, and you associate that knowledge with a familiar kitchen item like a stove. Your next memory is stored in a bedroom, and you imagine some characteristic element of the bedroom, such as a bed with blue sheets. The process continues until you have "stored" each piece of knowledge within a different part of an imaginary house, and your memory recall is strengthened and stimulated by your episodic recollection of moving around the house. This technique to strengthen your episodic memory connects otherwise abstract or meaningless data to the common task of walking around a house. Also, this exercise introduces an artificial yet effective coherence and episodic, continuous quality to the data you need to commit to memory.

When you imagine yourself walking through the imaginary house you constructed to hold the knowledge you want to memorize, you engage in a mental perception similar to Aristotle's understanding of memory. You are prompting your mind to engage as if your eyes perceive a tangible house with rooms and distinctive furniture, yet this perception is artificially constructed within your own mind. Nonetheless, perception and "viewing" the rooms and their respective memories is an unmistakable characteristic that confirms Aristotle's point of view.

How did the philosopher Henri Bergson explain how memories interact with the brain? What is the metaphysical worldview of Henri Bergson regarding mind-brain duality?

In his 1896 book, Matter and Memory, philosopher Henri Bergson (1859 - 1941) believed that while habitual memory or "muscle memory" for routine tasks may be inscribed within the brain, episodic or "pure" memory is not within the brain, or even within our current time and space. Instead, Bergson applied a time-based philosophy to memory and consciousness. While Rene Descartes divided mind and matter into dual, opposing substances, Bergson divided the mind or spirit into the past while the body occupies the present.

What is the time-based theory of memory proposed by Henri Bergson?

The temporal or time-based theory of memory proposed by Henri Bergson anchors the mind (or spirit, or soul) into the past, and the physical body and brain are always stuck in the present. Our perception and action of consciousness involve consideration of the past, and how memories may influence our present, physical actions that spill forward into the future. For Bergson, the more you consider your memories, the more you exercise consciousness. If you instead "live in the present," totally within your physical body, only responding to stimulus or emotion or immediate distractions, you abandon higher consciousness and approach the status of an unthinking animal. Past, present, and future integrate and reveal themselves through a unity of the mind and body, as consciousness (mind, spirit) evaluates the past to engage the present (through the brain and body) and influence the future.

How does the understanding of memory outlined by philosopher Alfred North Whitehead overlap with the worldview of Henri Bergson?

Philosopher and mathematician Alfred North Whitehead (1861 - 1947) believed that recalling memories involves engagement with the past. Rather than accessing some physical location in the brain as expected by materialism, Whitehead believed that the mind connects with the past in some immaterial process to access memories. To Whitehead, memory recall involved "prehension," a form of reaching back into and grasping those events from the past, rather than physically retrieving the memory like a computer.

What is the controversial and provocative theory of morphic resonance?

Scientist and philosopher Rupert Sheldrake (born 1942) proposed the controversial and provocative idea of morphic resonance, whereby past patterns of activity and form influence future patterns of activity and form, across time and space. The theory of morphic resonance explains how animal behavior is passed on through non-genetic transmission: organisms are "tuned in" to the influence of prior behavior by similar species. For example, the spider knows how to spin a web through influence (memory through morphic resonance) from the past of millions of spiders spinning past webs. Each species of organism has its own habit or memory of both form (body shape, growth, development) and behavior (like instinctual activity) that is inherited collectively from millions of previous organisms. By the morphic resonance theory of memory, our individual memories differ in degree, but not kind, from the collective memory of form and behavior shared across entire species, including humanity.

Sheldrake has proposed an entirely novel worldview for memory wherein living creatures share in communal memory. Instead of the brain physically storing memories, under morphic resonance, every person is "tuned into" and influenced by the memories and past behavior of organisms most closely related. The person most closely related to you is yourself, which is why you can recall your memories by their influence from your past, in the same way that behavior and form is transmitted collectively across time and space through morphic resonance.

Morphic resonance is an incredibly sweeping theory and has specific applications to human memory. If memory is inherent in all nature, then memories are accessible from somewhere else, or through some other medium than the physical brain. While morphic resonance remains controversial, the implication that memory is not stored in the brain does align with decades of neuroscientific research that has failed to identify a physical location of memories in the brain. For more information about morphic resonance, research the work of Rupert Sheldrake in Morphic Resonance: The Nature of Formative Causation (2009, originally proposed in 1981 as A New Science of Life: The Hypothesis of Morphic Resonance) and Science Set Free (2012, titled The Science Delusion in the United Kingdom edition).

33 Near-Death Experiences

Near-Death Experiences provide compelling evidence that the immaterial mind functions without the physical brain. While in various states of severe physical trauma or even clinical death, including complete brain death and heart stoppage, thousands have reported the memory, experience, and perception of reaching the afterlife and returning. During this strange journey to the brink of death, the mind, soul, or consciousness leaves the body yet persists. Without the physical use of the brain, eyes, or ears, these Near-Death Experiences nonetheless involve incredibly vivid perception, experience, communication, sights, sounds, and other mental activities. None of this mental activity is remotely possible within atheistic materialism, nor within the worldview that the mind is nothing but the physical brain.

Even more shockingly, scores of Near-Death Experiences involve the objective verification of observations that were impossible for the patient to physically achieve while dying in the emergency room, but were nonetheless reported as viewed or heard through the out-of-body experience of the immaterial mind (or soul) beyond the brain. Due to the technological progress of emergency medical intervention, there has been an explosion of documented Near-Death Experiences involving credible witnesses and corroborators from reputable professions, including the astonished medical doctors who listen to their patients describe in detail how the surgery proceeded from an immaterial, out-of-body vantage point of several feet above the patient's own body. The atheistic materialist worldview is forced into the untenable position of declaring that every single Near-Death Experience is a hoax or false memory.

What is a Near-Death Experience?

A Near-Death Experience is when you come close to the point of death from serious injury, disease, or life-threatening trauma but instead return to life after perceiving both the natural environment surrounding the body and also experiencing a journey to a different and somehow more vivid reality. Typically, a Near-Death Experience starts by perceiving your consciousness or sense of self leaving the physical body, allowing you to see and hear your own body and its surrounding persons, objects, and setting, without using the eyes, ears, or sensory organs of the body. After leaving your own body, you travel through a tunnel or portal to a place unmistakably different from our natural world, where you encounter and communicate with other beings. These experiences are uniformly described as clear and lucid as distinguished from murky and disordered hallucinations or dreams.

Have Near-Death Experiences been extensively studied, analyzed, and explored by academic researchers?

Yes. Thousands of documented Near-Death Experiences have been studied and analyzed by scientific and academic researchers. Cardiologists, neurosurgeons, psychologists, philosophers, theologians, and many other academics and practitioners have encountered or studied these fascinating events that unfold mysteries of the mind, brain, life, and death. Numerous books have been published about Near-Death Experiences, and the topic is covered extensively in prestigious, peer-reviewed, academic, scientific, and medical journals.

What are the most common and corroborated features of Near-Death Experiences? What distinguishes Near-Death Experiences from mere dreams or hallucinations? What places and persons are encountered during Near-Death Experiences?

First, Near-Death Experiences are incredibly vivid, lucid, and clear, and often described as more realistic than natural reality, which precludes hallucination, anesthesia-induced delirium, lack of oxygen, or other mistaken perceptions. Most Near-Death Experiences begin by leaving the physical body, whereby your consciousness, mind, soul, or self-perception floats up into the air, in a straight path away from the body. The most common initial perception is the ability to see your own body as "you" float above your body, in some intangible state of being. Another common experience is being able to view and hear people and events happening around your prone body, as your point of view shifts to an overhead vantage.

Next, the common experience is entering through a tunnel or passageway to a different world or reality that seems incredibly vivid and detailed, far beyond the clarity viewed in our world. At the other side of this tunnel is usually a very pleasant, beautiful, peaceful world (like Heaven), but in a minority of experiences, the other side is unpleasant, frightening, and violent (like Hell). In many experiences, the person undergoes a "life review" or abbreviated sequential summary of important actions or failures to act in that person's life, usually focused on the concept of loving others rather than any material, financial, professional, or artistic accomplishments.

Do Near-Death Experiences involve contact with deceased friends and relatives? Does anyone meet God or angelic beings during Near-Death Experiences?

Yes. Many Near-Death Experiences involve interaction and communication with divine, spiritual figures, perceived to be God or angels. Frequently these experiences also include meeting dead relatives or friends but never encountering anyone who remains alive on earth. The most frequent message from these otherworldly entities is that the time of death has not been reached, and the person must return to life, thereby completing the Near-Death Experience. Upon returning to the physical body, going through the Near-Death Experience often changes profoundly the way the person lives out the remainder of life.

How do the most common and corroborated features of Near-Death Experiences compare to traditional religious beliefs?

The obvious interpretation of the common features of Near-Death Experiences aligns remarkably well with traditional religious beliefs. The soul or spiritual essence leaves the body, but the individual consciousness and sense of subjective perception remain with the immaterial soul or spirit, not with the physical body. The soul, separated from the body, travels through a portal to the spiritual realm, which reveals a pleasant, beautiful Heaven or a frightening, nightmarish Hell. Finally, the encounter with God or angels, or dead relatives in the company of the divine, aligns with the religious understanding of the afterlife or destination for the dead.

What do Near-Death Experiences suggest about the mind-brain relationship?

Near-Death Experiences suggest that the mind is much more than the brain, that the mind is not merely the product of brain activity, and that the mind can survive and operate without the brain. Near-Death Experiences confirm mind-brain dualism and contradict atheistic materialism.

What do Near-Death Experiences suggest about consciousness?

Near-Death Experiences involve consciousness independent of the brain and the body, thereby conflicting with materialism that considers consciousness to be a byproduct of physical processes in the brain and body. Atheism is also refuted by the survival of consciousness after death within an "afterlife" or alternate realm of existence involving God and divine beings. Therefore, Near-Death Experiences pose serious challenges to the atheistic materialist worldview.

How do some Near-Death Experiences provide evidence that the mind is more than the product of the brain, and that the mind can exist and function without the brain?

During some well-documented Near-Death Experiences, the person is also experiencing brain death, when the brain has no activity. Yet the subject returns to life, remembers, recalls, and recounts the Near-Death Experience, which should not be possible if those mental activities are completely dependent on the brain.

Even more intriguing is the fact that in many Near-Death Experiences the person returns to life with knowledge of events and details that were observed or perceived in a manner impossible under materialist explanations. For example, after leaving the body, the person watches from above as surgeons and nurses perform the operation.

After returning to life, the person remembers and shares the precise actions, sounds, music, surgical instruments, and procedures that did, in fact, happen, like in the case of Pam Reynolds, discussed below. But there should be no way to perceive any of those facts because the surgery or trauma patient has eyes taped shut and is experiencing brain death or heart stoppage, commonly called "flatlining." Under the materialist worldview, the person should not be able to perceive or remember anything, which challenges the idea that the mind is nothing but the product of the brain.

How many Near-Death Experiences have been documented throughout human history?

Tens of thousands of Near-Death Experiences have been documented. While the number of recorded Near-Death Experiences is significant, the actual number of Near-Death Experiences is almost certainly many orders of magnitude larger, perhaps tens of millions just within modern times. It is likely that the overwhelming majority of Near-Death Experiences were not recorded for posterity but instead lost to the memory of time. There are scientific journals devoted entirely to Near-Death Experiences.

Why have Near-Death Experiences become more commonly experienced and studied?

Put simply, we now have better technology to bring people back from the brink of death and to document the process. Advances in medicine combined with increased research interest has drastically increased the number of verified and documented Near-Death Experiences. With incredible technology to revive patients in trauma centers and emergency rooms, and sophisticated medical devices to monitor and record vital signs, conditions are ripe for capturing objective evidence to corroborate Near-Death Experiences.

What is the one of the most famous Near-Death Experience that provides evidence of the power of the mind beyond the brain?

Pam Reynolds (1956 - 2010) suffered a severe brain aneurysm in 1991. The blood vessel in her brain was going to burst, and neurosurgeon Robert Spetzler (born 1944) of the Barrow Neurological Institute performed an extraordinary intervention. Spetzler brought Reynolds to a condition of clinical death by cooling her body to 60 degrees Fahrenheit, placing her on a heart and lung machine, stopping her heartbeat, and draining all the blood from her brain by tilting the body. Spetzler also inserted into Reynolds's ears molded earplugs that played incredibly loud clicking noises every second. The purpose of the loud noises was to monitor when Reynolds' brain activity stopped reacting to the loud noises, which indicated brain death and the opportunity to begin surgery. After forcing Reynolds into this precarious state of physical death, Spetzler healed the aneurysm.

To stress the point, the elaborate procedure took about six hours, and for significant periods, Reynolds was clinically dead, with no heartbeat, and no brain activity (nor blood in the brain). Spetzler removed the aneurysm and then restarted Reynolds' heart and lungs, and amazingly, Reynolds recovered and returned to life.

Upon waking from the procedure and being asked by Spetzler about her recovery, astonishingly, Reynolds volunteered that she witnessed and heard the entire operation. She was able to correctly describe the surgical instruments, details of the procedures, the music and song playing in the operating room, and the other assistants helping Spetzler with the surgery. Reynolds reported that she saw and heard all these details while she was floating above her clinically dead body in the operating room. Recall that Reynolds had her ears plugged and was subjected to extremely loud noises every second, and that the brain scans showed no activity or response. Also, Reynolds had her eyes taped shut, in addition to being clinically dead according to "flat scans" for both her heartbeat and brain activity. When you combine these facts with a blood-drained brain and a body cooled to 60 degrees Fahrenheit, there is no question that under the materialist worldview, there was no way for Reynolds to be conscious at all, let alone to hear or see or remember anything.

Reynolds also relayed the commonly corroborated details of a vivid and wondrous Near-Death Experience. She went through a tunnel to another realm where she encountered dead relatives who instructed Reynolds to return to life. As she experienced the return to her body, she felt as though she dove into a pool of ice water, which corresponds to her body being cooled to 60 degrees for the procedure. During the entire time that Reynolds perceived and remembered details of the operation, her eyes were taped shut, her ears were sealed off and subject to extremely loud noises, her brain was drained of blood, and her lack of vital signs meant that she was clinically dead.

The case of Pam Reynolds is just one of many that present evidence of the immaterial mind existing, perceiving, and functioning without input from the physical brain. These Near-Death Experiences directly contradict the materialist conception of the mind as nothing but the product of the brain.

What are some other anomalous results of Near-Death Experiences that challenge the atheistic materialist worldview, and the reductive understanding of the mind-brain interaction? Why do these aspects of Near-Death Experiences disprove atheistic materialism?

The following is a summary of hundreds of Near-Death Experiences that indicate consciousness, perception, memory, thought, movement, and communication during out-of-body experiences of the immaterial mind rather than the brain. Each of these descriptions invalidates the atheistic materialist worldview and can be cross-referenced using the many books, academic articles, research studies, and databases devoted to Near-Death Experiences:

Ability to see and hear without using the eyes or ears of the body.

Ability of blind people to see and of deaf people to hear during the experience.

Ability of blind people to recall and describe visual experiences upon returning to life.

Ability of deaf people to recall and describe auditory experiences upon returning to life.

Consciousness, the mind, the self, the soul experienced away from the brain and body.

Ability to remember and recall experiences from outside the body upon returning to the body.

Ability to think, reason, contemplate, and make decisions without the brain or body.

Ability to perceive and feel emotions and intentions without the brain or body.

Ability for the self apart from the body to move in all directions subject to no gravitational force, at superhuman speed, and to pass through solid objects like doors.

Communication by a type of inner telepathy, rather than by verbal speech and auditory hearing, with divine entities identified as God and angels.

Communication with the dead or with the souls or spirits of the dead.

None of these experiences, abilities, or perceptions should be possible without the brain and body according to the atheistic materialist worldview. The obvious implication of this data is that the immaterial mind, personal consciousness, and immortal soul exist beyond the material body and brain and that human experience is not bound by the material.

Where can you learn more about Near-Death Experiences?

Near-Death Experiences have been extensively studied for decades in scientific, scholarly, peer-reviewed journals. The *Journal of Near-Death Studies* is exclusively dedicated to the topic, publishing peer-reviewed articles since 1982. Helpful books include After: A Doctor Explores What Near-Death Experiences Reveal about Life and Beyond (2021) by psychiatrist and scholar Bruce Greyson (born 1946), Imagine Heaven (2015) by pastor and theologian John Burke (born 1962), and The Handbook of Near-Death Experiences: Thirty Years of Investigation (2009, various editors). The first book to coin the term, "Near-Death Experience," is by psychiatrist and philosopher Raymond Moody's (born 1944) pioneering work, Life After Life (1975).

34 Mind Over Matter: The Placebo Effect

The placebo effect is one of the most scientifically studied and confirmed connections between the immaterial mind and physical body. Your thoughts, feelings, and intentions can and do affect your medical health, recovery, and response to disease. By introducing to your immaterial mind the thought of recovery and healing, the placebo effect somehow affects the physical body in a positive way, supercharging the healing process. Atheistic materialism cannot explain a purely physical mechanism for the placebo effect. Therefore, the placebo effect is one more reason to adopt a worldview that considers immaterial causes, including the immaterial mind beyond the physical brain.

What is a placebo?

A placebo is a medical treatment, dosage, or intervention that has no therapeutic effect according to conventional medicine. For example, the most familiar placebo is a pill to swallow that looks like a prescription pharmaceutical drug but is actually just sugar. Since the pill is only sugar, there are no active or therapeutic drugs that you would expect in a prescription pharmaceutical product. Placebos can also be other treatments, rituals, interventions, or any other procedures, as long as there is no known active ingredient or conventional therapeutic mechanism.

Does taking a placebo have a powerful, measurable effect on health, recovery, and response to disease and injury?

Surprisingly, yes. One of the best-established facts of medical research is that placebos do, in fact, create positive change in those who ingest or partake in them. This mysterious property of placebos is so famous that it has a name: the placebo effect.

What is the placebo effect?

The placebo effect refers to the mysterious and well-documented phenomenon of patients responding with positive improvements to interventions that have no active ingredients or therapeutic mechanisms—like sugar pills. Even though the placebo, in whatever form, has no medical or pharmaceutical mechanism for healing, positive response, or health improvement, nonetheless, the patient improves.

What is one proof of the recognized power of the placebo effect?

The placebo effect is so powerful that it is standard practice to test the efficacy of new pharmaceutical drugs against the placebo effect. Researchers and regulators recognize that the placebo effect is real and effective so new drugs must demonstrate superior results to the placebo effect to pass scrutiny.

What are some explanations for how the placebo effect works?

The placebo effect suggests that non-material forces can effect material change, specifically in health outcomes. While traditional medicine is mechanical, chemical, and physical, the placebo effect is purely the result of a change in the state of immaterial mind. By merely thinking or believing that an otherwise inactive substance or process will improve physical health, the immaterial change (different thoughts or beliefs) directly affects material conditions (health of the physical body). Therefore, the placebo effect directly challenges the materialist, reductive, mechanistic theories of the mind-brain interaction. In the battle between mind over matter, the placebo effect is an unquestionable victory of the immaterial mind.

35 The Limits of Computer Intelligence and the Chinese Room Argument

Materialists believe the mind is nothing but the physical brain, and the brain is a machine like a computer. For these reasons, some believe that computers can match or surpass the mental capacity of human beings, or even produce a sentient, conscious, intelligent life form from software code. These false beliefs come from a misunderstanding of the nature of the human mind and its special properties of intentionality, subjectivity, experience of qualities, free will, and understanding as outlined by philosopher J.P. Moreland. While computer programs can imitate and simulate thinking, speaking, and decisions, the famous Chinese Room Argument debunks the misconception that computers do anything other than deterministically follow programs written by human beings.

What distinguishes the human mind from computer functions, like artificial intelligence programs? According to philosopher J.P. Moreland, what are five aspects of the human mind that artificial intelligence can never replicate?

Philosopher J.P. Moreland (born 1948) proposes five aspects of the human mind that artificial intelligence (and computer programs) can never replicate: intentionality, subjectivity, qualia, free will, and the ability to understand. While sophisticated computer programs can simulate or create a false impression of these five activities, so-called artificial intelligence is not truly performing any of them.

How do the five aspects proposed by J.P. Moreland—intentionality, subjectivity, qualia, free will, and understanding—help to explain the concept of consciousness and the mind?

Our minds and consciousness exhibit many properties and perform many actions. Minds manifest intentions and purposes about other objects or goals. Consciousness involves distinct subjectivity, the first-person perspective, or the sense of self. Our state of consciousness includes experiencing qualia, sensations, or feelings that cannot be quantified and are inherently subjective and personal. Our minds exercise free will to make decisions and employ the intellect to understand abstract concepts. These five aspects proposed by J.P. Moreland distinguish the mind and consciousness from routine computing functions, deterministic software code, and machine operations, none of which can accomplish any of the five aspects of the immaterial mind.

What is the categorical error committed by those who assign undue praise to computer programs?

The classic mistake is to confuse imitation for the genuine article, or to mistake simulation for the reality being simulated. Video games simulate all kinds of real-life activities, but we know they are not real. Money machines dispense currency and respond to requests, but we realize they only simulate the function of bank tellers. Artificial intelligence software is specifically designed to imitate or simulate human interaction because those functions appeal to human customers. But do not mistake your reflection in the pond for yourself.

Unending, overhyped press releases shout from the rooftops that so-called artificial intelligence software has reached yet another milestone, matching its intelligence or capabilities to that of humanity. The problem is that almost none of these claims are true.

Can computers think about anything? Can computers feel or have experiences? Can computers make conscious decisions? Can computers truly understand something?

No. Computers do not and cannot think about anything, feel or experience perception, make conscious decisions, or understand anything at all. What computers can do is merely simulate these processes following programs created by intelligent human beings. The intelligence comes from the software engineers, not from the engineered software.

Why is it impossible for computer programs or artificial intelligence to achieve the properties of the human mind? What are the inherent limitations of computers?

Computer programs only follow software code, applying an algorithm (set of instructions) to a given input, resulting in a predetermined output. No computer can perform anything other than what is written in its programming, which precludes anything remotely approaching free will. Even artificial intelligence programs that use large language models to simulate speech and conversation are not actually holding a discussion, but rather, simulating discussion. Based on your written or spoken prompts, the computer program searches a vast database of existing words and writings. Then, the program selects a string of words based on grammar and syntax algorithms combined with a probability equation for the next likely words or phrases.

You may be tricked into believing that the computer program is thinking, but in fact, the computer program is merely processing a program written by human intelligence that scans a large database of the products of human intelligence (books, websites, literature, articles, transcripts, etc.).

What is intentionality? Can a computer exercise intentionality?

Intentionality is your ability to think "about" something, or to focus your mind on a specific goal or topic. You can think about your next meal, or you could think about how you will pay for your next meal, or you could think about where you will eat your next meal. Each of these intentions offers a different direction or focus of your intentionality. Computers can only follow instructions in software code, and cannot exhibit intentionality, which requires subjective exercise of free will and abstract thought directed toward a specific topic.

What is subjectivity? Can a computer exercise subjectivity?

Subjectivity refers to your first-person perspective, the fact and experience of your awareness of your own being. Computers are material machines that respond to physical stimuli. Just like your car responding to your foot on the gas pedal, computers respond to input according to their software code. No matter the sophistication of the computer programming, or how closely the computer output mimics human speech, a computer system has no first-person perspective, nor can it experience the sense of its own being.

What are qualia? Can a computer experience qualia?

Qualia are the subjective, qualitative experiences of your sensory perception. The color orange, the smell of sulfur, and the pain of biting your tongue are all qualia that cannot be fully explained or understood without direct experience. The subjective experience (qualia) of the same object may be different between persons and is often different when experienced by other animals. For example, the range of vision for some animals includes portions of the light spectrum not detectable by human eyes. Computers share none of the complex biological systems of human beings, nor do they have a mind, nor a first-person perspective, nor qualitative experiences at all.

Do computers actually understand anything?

No. Computers do not understand anything. Computers follow software code and respond to stimulus according to deterministic algorithms. No matter the complexity of the computer program, nothing is understood, and everything is constrained by instructions (software code). Computers are tools like calculators, hammers, or typewriters. While computers may deceptively simulate understanding, any such simulation reflects the intelligence infused by human beings into the programing code. One of the best illustrations of this principle is the Chinese Room Argument, first proposed by philosopher John Searle.

What is the Chinese Room Argument?

American philosopher John Searle (born 1932) introduced the Chinese Room Argument in a 1980 article, explaining how the appearance or simulation of understanding is different from true understanding. Imagine you find yourself inside a room with many boxes of cards written in Chinese and a thick book of instructions written in English. The cards have Chinese-language messages written on the front and back.

Outside the room, Chinese speakers pass you cards under the door that have questions written in Chinese. You know only the English language. You do not read, write, speak, or understand Chinese. Yet by reading the English instructional book, for hours, you successfully return cards from the boxes to the Chinese speakers outside the room. Based on what you see on the cards they present, you read the instructional book written in English, which directs you to hand out specific cards written in Chinese. Each card you return to the questioners has the correct answer, written in Chinese. The Chinese speakers outside the room are utterly—and mistakenly—convinced that someone inside the room understands Chinese.

How does the Chinese Room Argument explain the misconception that computers understand anything?

In the Chinese Room Argument, your performance is that of a computer program, perhaps software code for so-called artificial intelligence to simulate a conversation. The cards slipped into the room represent the input or prompts that you present to the computer program. The instructional book written in English is the software code or algorithm. The box of cards is the computer database, just like large language models might scan the entire internet or libraries of published books and articles. The cards you pass under the door to respond would be the program output or the simulated "answers" or "conversation" of sophisticated artificial intelligence designed for simulated conversation.

Just like you do not understand Chinese, even after passing the cards back and forth, the computer program does not understand anything or any conversation. Even as you fool the Chinese speakers into believing that someone inside the room is a Chinese speaker, sophisticated software can trick you into perceiving some type of intelligence or understanding within the computer program—especially within a simulated "conversation." Yet in both cases, there is no understanding, only simulation thereof by carefully following a set of predetermined instructions.

36 Technocracy, Transhumanism, and Artificial Intelligence

So-called "artificial intelligence" is closely linked to transhumanism. Technocrats want people to become more like machines and want machines to become more like people. But these dark fantasies reveal philosophical and metaphysical weaknesses: immaterial minds are fundamentally different from physical machines, and the human mind cannot be replicated by steel or software, no matter how ingeniously designed. Ironically, atheistic materialists deny the spark of life, soul, consciousness, and immaterial mind in humanity, yet believe they can summon all these precious gifts from lifeless machines.

What is technocracy? What is a technocrat?

Technocracy means "rule by technology." The corresponding political worldview is that we should be ruled by the elites of technology and industry: scientists, engineers, computer and software experts, and especially titans of industry like the CEOs and owners of technology companies.

What is the typical worldview of a technocrat?

Most technocrats are atheistic materialists who do not believe in God, souls, or spirits, and who believe the human mind is reducible to the physical brain tissue in the skull. Many believe that humans are nothing more than machines, and therefore, computer machines can duplicate and eventually replace most human functions. An especially strange subset of technocrats believe in transhumanism.

What is transhumanism? What is the origin of the term? What is the goal of transhumanists?

Transhumanism means to "change humanity" and the desired change is to come from technological enhancement: genetic engineering, cybernetic implants, artificial intelligence, nanotechnology. By merging the human body with machines and computers, transhumanists believe a new form of humanity will emerge, beyond our current capabilities and existence. Notable atheist, eugenicist, and evolutionary biologist Julian Huxley (1887 - 1975, grandson of Thomas Henry Huxley, "Darwin's Bulldog" or strongest contemporary advocate of Charles Darwin) coined the term in his 1957 essay, "Transhumanism."

Many transhumanists believe that technology-enhanced people will "transcend" and "transform" into a superior form of life. Artificial intelligence is one of many technologies that transhumanists seek to merge with the human mind and body to reach this transcendent state.

What is the greatest deception of so-called artificial intelligence?

What is called artificial intelligence is not intelligence, but it is artificial. The term itself is inherently deceptive. Just about anything described as artificial intelligence is actually some combination of machine learning, computer programming, and algorithmic parsing of large language models.

Why should you be skeptical about software programs being described as artificial intelligence?

Software programs written by human beings do not magically assume the property and existential condition of being intelligent, even if the program deceptively simulates limited aspects of human cognition, decision-making, or pattern recognition.

What novel computer functions in the early 21st century led to the deceptive use of the term artificial intelligence?

Before computer software programs involved concepts like machine learning, predictive text outputs, large language models, and generative chat-bots, nobody serious made any claims about artificial intelligence. But now that software programs simulate aspects of human activity, like speech, writing, and pattern recognition, the technology marketers and advertising pitchmen gush and brag about breakthroughs in so-called artificial intelligence.

What is the proper, original meaning of artificial intelligence?

Originally, artificial intelligence meant what is now described as "strong artificial intelligence" or "general artificial intelligence." The term artificial intelligence has been diluted and abused to include mere software programs.

Before software nerds, blithe journalists, and carnival barkers of Silicon Valley robbed the term of all meaning, artificial intelligence meant a system that possesses human-level cognitive abilities, or even a sentient lifeform. Yet now, the ability for machines to think, learn, reason, and adapt across a wide variety of functions and domains has shifted to the terms "strong artificial intelligence" or "general artificial intelligence." Be careful with this semantical bait-and-switch. What is called artificial intelligence is nothing more than computer programming trained on large data sets to recognize patterns and select probabilistic connections.

Does "strong artificial intelligence" or "general artificial intelligence" exist?

No. It is extremely unlikely or impossible that "strong artificial intelligence" or "general artificial intelligence" will ever exist. When you read a sensational news article that announces either goal has been achieved, ask yourself whether you believe that the dazzling performance of this computer program is equal to the power of the human mind.

What is the semantical trick used to make claims about artificial intelligence rivaling, meeting, or surpassing the power of the human mind?

If the human mind is reduced to a computer, then artificial intelligence—even "strong artificial intelligence"—has already been achieved. Of course, the mind is much more powerful than any computer program. The semantical trick used by artificial intelligence hucksters is to reduce and diminish humanity to magnify and embellish the feats of computers. Achieving parity with the human mind gets easier the more you lower the bar, and the more limited and constricted view you adopt of the human mind. If the human mind is nothing but a machine, then of course, machines can surpass humans.

What is a memorable example of the deceptive nature of claims that artificial intelligence or software coding has surpassed the human mind?

Unsurprisingly, computer programs can reliably beat the best chess players, and even the best go players. (Go is a Chinese strategy game with even greater complexity and orders of magnitude more combinations of possible moves than chess.) Predictably, artificial intelligence evangelists bragged that software programs surpassed the human mind. Yet the deceptive trick is evident for anyone who examines the bold claim. Playing chess (or go) is not the summit and crown of achievement of the human mind. These games are just one of infinite possible activities available to the wondrous human mind. Furthermore, no computer program has ever played chess.

Why has no computer program ever actually played chess?

No computer program has ever played chess. Instead, computer programs have ran sequential algorithms that search through all possible chess moves (according to the rules) and calculate the odds of winning the match based on the next selected movement of a chess piece. Calculating odds, considering mathematical probabilities, and selecting the optimal next move is not the same as playing a game of chess. Computer programs do not understand anything about chess or strategy, they simply calculate based on software programs written by human beings.

How does describing as artificial intelligence the software code that calculates the next chess move commit a deceptive logical fallacy?

While the computer program can simulate the activity of playing a game of chess, the computer program does not perceive the opponent across the table. The computer program does not experience a game of chess, nor does it think "about" the next move. The computer program does not think at all, nor does it win or lose, nor become stressed or flustered or emboldened. All the computer program does is calculate the next move based on a mathematical, probabilistic determination among all possible next moves, constrained by the rules of chess written into software code.

Yet all these other activities, experiences, and conditions—and much more—are required elements of playing chess. Software programs are successful in simulating a wide range of activities, but that is different from experiencing the fulsome range of subjective, personal activities. Describing the simulation of an intelligent, sentient activity (like playing chess) as artificial intelligence that surpasses human intelligence is deceptive and illogical. Not only did the computer not perform the same human activity, but the measurement of man is not confined to a board game.

How is the desire to create artificial intelligence an ancient form of wish fulfillment?

The race to create artificial intelligence through computer software programs is only the latest manifestation of the ancient human desire to create life. Some portion of mankind has always strived to assume the throne of God—to call himself the creator rather than to recognize his station as the created. See chapter 19, "The Faustian Bargain: Theocentric vs. Anthropocentric Worldviews."

What is the assumed process by which artificial intelligence could emerge from a computer program?

The murky, minimal details of the process for the emergence of artificial intelligence from computer programming leave much to be desired. Some believe that as pattern-recognizing (so-called "predictive") software code becomes more sophisticated, the computer program itself will gain consciousness. Exactly how this life and subjective consciousness will emerge is rarely explored, rather, it is thrust upon the world as a bold assumption or promise. While some place faith in the increasing sophistication of software, others look to the expansion of available data to process.

How does the philosophical worldview of the technocrats delude them into believing the fantasy of creating life, intelligence, and consciousness from computer programs or machines?

If humans are nothing but machines, as many technocrats believe, then they should be able to build a superior version of humanity. Believing that machines can become alive, intelligent, or surpass humanity is an inherently materialist, reductive worldview. The same technocrats who believe that they can create life and consciousness also believe that your consciousness is nothing more than a byproduct of the brain tissue in your skull (eliminative materialism). They also believe your life is a strictly material phenomenon, with no inherent meaning, purpose, or value. Rather than being a gift from God, your very soul or spirit is a primitive superstition (the classic worldview of atheistic materialism).

Further, they believe that the best metaphor for your brain and your mind is a computer (reductive, mechanical materialism). Some even drop the metaphor and would profess that your brain truly is a computer, consisting of organic tissue rather than inorganic machine parts, and that your consciousness is an illusory byproduct of the brain (the materialist identity theory of the mind-brain interaction). All these materialist worldviews deny the existence or importance of the immaterial mind and soul, reducing everything to material objects, such as the brain.

How does the mastery of computer manufacturing and software coding, combined with an impoverished understanding of humanity, contribute to the fantasy of creating new life in the form of artificial intelligence?

These technocrats have mastered the process of manufacturing computers, and of writing software code. If your mind is just like a software program, and your brain is simply another version of computer hardware, they believe they can reproduce the human mind as easily as they can invent the next cutting-edge laptop computer or sophisticated software code. If you are nothing but a sophisticated meat-machine, the technocrats believe they can manufacture an updated version of humanity.

What is the fundamental category error of believing that life can emerge from machines? What is the metaphysical irony of hoping for consciousness to emerge from machines?

Machines are not alive, nor can machines ever become alive. The breath of life cannot be breathed, by mankind, into machines. The metaphysical irony is that the same people who deny that consciousness is anything but the brain believe they can create a new form of consciousness using machines, computers, and software. Atheistic materialism leads to absurd conclusions, like denying fundamental aspects of humanity like the soul, the mind, and consciousness, but simultaneously believing those immaterial elements can be summoned from lifeless machines. Only through some unholy combination of hubris, philosophical ignorance, and metaphysical delusion could someone think it possible to create a new life form from cold computer hardware or lifeless software code.

37 Concluding Thoughts

What did we learn in Part 1, the first five introductory chapters to this book?

Scientism is the mistaken belief that no truth exists except that which is proven by science. Scientism is a false and incomplete worldview because the practice of science requires assumptions and beliefs not proven by science. Science emerged from the religious worldview of European Christendom because of the belief in a rational, ordered universe intelligible to mankind created in the image of God. Therefore, science is not inherently opposed to religious faith, and the scientific outlook rests upon decidedly philosophic and theological positions. While the sterile worldview of atheistic materialism limits our thinking to exclusively material causes, we have compelling evidence for the immateriality of the universe and our minds.

What did we learn about the immateriality of our universe within Part 2, Matter and Cosmos?

The Big Bang: All matter, energy, time, and space appeared suddenly some 13.8 billion years ago from a state of absolute nothingness. Therefore, materialism cannot explain the origin of the cosmos because there was no matter to provide a material cause.

Fine-tuning: Our cosmos and the laws of physics are incredibly fine-tuned in myriad ways, signaling that the universe did not arise by chance or necessity.

Quantum mechanics: Even solid matter exhibits immaterial properties at the quantum scale. Quantum physics points to an immaterial, abstract nature of reality that is intelligible only to immaterial minds like ours.

What did we learn about the immateriality of our minds within Part 3, Mind and Consciousness?

Consciousness, free will, and the immaterial mind provide insurmountable obstacles to the atheistic materialist worldview that asserts the mind is nothing but the physical brain. Studying consciousness, free will, memory, and even the brain itself reveals the immaterial nature of the human mind, which cannot be replaced by so-called artificial intelligence or computer programs. Our immaterial minds refute the atheistic materialist worldview because we are more than mere machines or meat puppets. Our lives have meaning, purpose, and value that supersede materialist dogma, and our minds have immaterial properties beyond the physical brain.

How is your life changed by the exercise of your free will through conscious decisions about how you view the nature of reality?

You are more than moving meat; your mind is more than three pounds of organ tissue encased in your skull. Your life has infinite potential, and the end of your earthly life is the beginning of the next. You have an objective, transcendent purpose beyond any goal, vocation, or lifestyle you choose in your limited time on earth. The meaning of life is within your grasp, but only if you grant that the ultimate meaning is above you rather than beneath you.

If you stare and strain at the ground, with a craned neck and bent spine, you may get to the bottom of everything but never rise above the dirt under your feet. Reducing your existence and our cosmos to the tiniest level of detail will yield impressive calculations, astonishing descriptions, but nothing comprehensive, fulfilling, or transcendent. Mere objects in motion yield numerical values expressed in the language of mathematics, not virtues spoken in the Word of love.

If you believe in the tragic lie that you are nothing more than an evolved ape, and that you are shackled by the physical impulses of your brain and bodily organs, that self-fulfilling prophecy will engulf you. Slouching into your place as merely the most powerful brain in the animal kingdom, you will abdicate your throne and majesty. If you instead recognize your exalted status within mankind, made in the image of God, you can orient yourself to the source of all being, rather than losing yourself among the distractions of actions. Look up, not down, for your origins. Search outside your body, not within your skull, for the nature of your mind.

If you deny Heaven, Heaven will be denied. If you relinquish your soul, your soul will be relinquished. If you reject God, your wish will be granted. Free will is necessary for love, as love cannot be forced but only given freely. If you close your heart, your privacy will be respected, at the cost of community and communion. These choices have the greatest and gravest of consequences. Eternal, transcendent love is within your grasp, but you must believe in its power to enjoy its glory.

Searching for meaning, purpose, and value exclusively within yourself may grant you self-satisfaction and even exciting distraction, but love will remain outside you. By emptying yourself through self-sacrificial love for others, you will make room for an infusion of the greatest Love.

What should you do if you found this book intriguing, insightful, or helpful?

Please leave a fantastic, five-star review on Amazon and check out my other books.

What other books are available?

Christoph Mlinarchik is also the author of Christianity in Plain English, Government Contracts in Plain English, Federal Acquisition Regulation in Plain English, and Government Contracts Negotiation, Simplified. All the books are available in paperback and Kindle on Amazon.

The Government Contracts in Plain English Series:

https://www.amazon.com/dp/B09MRCMWBD?binding=paperback

Deep Questions in Plain English Series:

https://www.amazon.com/dp/B0FZWZ77HS?binding=paperback

How can you contact the author, Christoph Mlinarchik?

Email **Christoph@ChristophLLC.com** for any media, interview, podcast, public speaking, book-signing, discounted bulk book purchasing, or business inquiries.

Christoph Mlinarchik, JD, CFCM, PMP is an engaging and highly rated keynote speaker, available for conferences and other presentations. Many clients combine a box of books for attendees with a keynote speech followed by a book signing and Q&A session. Learn more at: **https://ChristophLLC.com**

Made in the USA
Middletown, DE
05 September 2026

32472710R00136